I0007346

Jewels
of *Stringology*

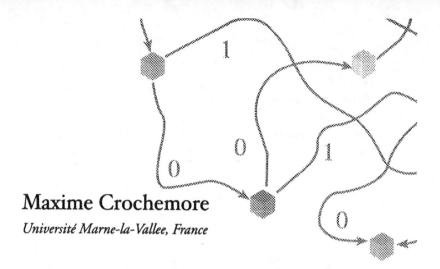

Maxime Crochemore

Université Marne-la-Vallee, France

Wojciech Rytter

Warsaw University, Poland & University of Liverpool, UK

Jewels
of
Stringology

World Scientific
New Jersey • London • Singapore • Hong Kong

Published by

World Scientific Publishing Co. Pte. Ltd.

5 Toh Tuck Link, Singapore 596224

USA office: 27 Warren Street, Suite 401-402, Hackensack, NJ 07601

UK office: 57 Shelton Street, Covent Garden, London WC2H 9HE

British Library Cataloguing-in-Publication Data
A catalogue record for this book is available from the British Library.

JEWELS OF STRINGOLOGY
Text Algorithms

Copyright © 2002 by World Scientific Publishing Co. Pte. Ltd.

All rights reserved. This book, or parts thereof, may not be reproduced in any form or by any means, electronic or mechanical, including photocopying, recording or any information storage and retrieval system now known or to be invented, without written permission from the publisher.

For photocopying of material in this volume, please pay a copying fee through the Copyright Clearance Center, Inc., 222 Rosewood Drive, Danvers, MA 01923, USA. In this case permission to photocopy is not required from the publisher.

ISBN-13 978-981-02-4782-9
ISBN-10 981-02-4782-6
ISBN-13 978-981-02-4897-0 (pbk)
ISBN-10 981-02-4897-0 (pbk)

Preface

The term *stringology* is a popular nickname for *string algorithms* as well as for *text algorithms*. Usually *text* and *string* have the same meaning. More formally, a text is a sequence of symbols. Text is one of the basic data types to carry information. This book is a collection of the most beautiful and at the same time very classical algorithms on strings. The selection has been done by the authors, and is rather personal, among so many famous algorithms that were natural candidates to be included and that belong to a field that has become now fairly popular.

One can partition algorithmic problems discussed in this book into practical and theoretical problems. Certainly string matching and data compression are in the first class, while most problems related to symmetries and repetitions are in the second. However, we believe that all the problems are interesting from an algorithmic point of view and enable the reader to appreciate the importance of combinatorics on words.

In most textbooks on algorithms and data structures the presentation of efficient algorithms on words is quite short as compared to issues in graph theory, sorting, searching, and some other areas. At the same time, there are many presentations of interesting algorithms on words accessible only in journals and in a form directed mainly at specialists. There are still not many books on text algorithms, especially the books which are oriented toward undergraduate and graduate students. In the book the difficult parts are indicated by a star, so the basic text becomes painless for undergraduate students. We hope that this book will cover a gap on algorithms on words in book literature for the broader audience, and bring together the many results presently dispersed in the masses of journal articles.

March 2002
M. Crochemore, W. Rytter

Contents

Chapter 1

Stringology

One of the simplest and natural types of information representation is by means of written texts. This type of data is characterized by the fact that it can be written down as a long sequence of characters. Such linear a sequence is called a *text*. The texts are central in "word processing" systems, which provide facilities for the manipulation of texts. Such systems usually process objects that are quite large. For example, this book probably contains more than a million characters. Text algorithms occur in many areas of science and information processing. Many text editors and programming languages have facilities for processing texts. In biology, text algorithms arise in the study of molecular sequences. The complexity of text algorithms is also one of the central and most studied problems in theoretical computer science. It could be said that it is the domain in which practice and theory are very close to each other.

The basic textual problem in stringology is called *pattern matching*. It is used to access information and, no doubt, at this moment many computers are solving this problem as a frequently used operation in some application system. Pattern matching is comparable in this sense to sorting, or to basic arithmetic operations.

Consider the problem of a reader of the French dictionary "Grand Larousse," who wants all entries related to the name "Marie-Curie-Sklodowska." This is an example of a pattern matching problem, or string matching. In this case, the name "Marie-Curie-Sklodowska" is the pattern. Generally we may want to find a string called a *pattern* of length m inside a text of length n, where n is greater than m. The pattern can be described in a more complex way to denote a set of strings and not just a single word. In many cases n is very large. In genetics the pattern can correspond to a gene that can be very long; in image processing, digitized images sent serially contain millions of characters each. The string-matching problem is the basic question considered in this book, together with its variations. String matching is also the basic subproblem in other algorithmic problems on texts. Following is a (not exclusive) list of basic groups of problems discussed in this book:

- variations on the string-matching problem

- problem related to the structures of the segments of a text

- data compression

- approximation problems

- finding regularities

- extensions to two-dimensional images

- extensions to trees

- optimal time-space implementations

- optimal parallel implementations.

The formal definition of string matching and many other problems is given in the next chapter. We now introduce some of them informally in the context of applications.

1.1 Text file facilities

The UNIX system uses text files for exchanging information as a main feature. The user can get information from the files and transform them through different existing commands. The tools often behave as filters that read their input once and produce the output simultaneously. These tools can easily be connected with each other, particularly through the pipelining facility. This often reduces the creation of new commands to a few lines of already existing commands.

One of these useful commands is *grep*, acronym of "general regular expression print." An example of the format of *grep* is

```
grep Marie-Curie-Sklodowska Grand-Larousse
```

provided "Grand-Larousse" is a file on your computer. The output of this command is the list of lines from the file that contains an occurrence of the name "Marie-Curie-Sklodowska." This is an instance of the **string-matching** problem. Another example with a more complex pattern can be

```
grep '^Chapter [0-9]' Book
```

to list the titles of a book assuming titles begin with "Chapter" followed by a number. In this case the pattern denotes a set of strings (even potentially infinite), and not simply one string. The notation to specify patterns is known as *regular expressions*. This is an instance of the **regular-expression-matching** problem.

The indispensable complement of *grep* is *sed* (stream editor). It is designed to transform its input. It can replace patterns of the input with specific strings. Regular expressions are also available with *sed*. But the editor contains an even more powerful notation. This allows, for example, the action on a line of the input text containing the same word twice. It can be applied to delete two consecutive occurrences of a same word in a text. This is simultaneously an example of the **repetition-finding** problem, **pattern-matching** problem and, more generally, the problem of finding **regularities in strings.**

The very helpful matching device based on regular expressions is omnipresent in the UNIX system. It can be used inside text editors such as *ed* and *vi*, and generally in almost all UNIX commands. The above tools, *grep* and *sed*, are based on this mechanism. There is even a programming language based on pattern-matching actions. It is the *awk* language, where the name *awk* comes from the initials of the authors, Aho, Weinberger, and Kernighan. A simple *awk* program is a sequence of pattern-action statements:

```
pattern1  {action 1}
pattern2  {action 2}
pattern3  {action 3}
```

The basic components of this program are patterns to be found inside the lines of the current file. When a pattern is found, the corresponding action is applied to the line. Therefore, several actions may be applied sequentially to a same line. This is an example of the **multi-pattern matching** problem. The language *awk* is meant for converting data from one form to another form, counting things, adding up numbers, and extracting information for reports. It contains an implicit input loop, and the pattern-action paradigm often eliminates control flow. This also frequently reduces the size of a program to a few statements. For instance, the following *awk* program prints the number of lines of the input that contain the word "abracadabra":

```
abracadabra  {count++}
END          {print count}
```

The pattern "END" matches the end of input file, so that the result is printed after the input has been entirely processed. The language contains attractive features that strengthen the simplicity of the pattern-matching mechanism, such as default initialization for variables, implicit declarations, and associative arrays providing arbitrary kinds of subscripts. All this makes *awk* a convenient tool for rapid prototyping. The *awk* language can be considered as a generalization of another UNIX tool, *lex*, aimed at producing lexical analyzers. The input of a *lex* program is the specification of a lexical analyzer by means of regular expressions (and a few other possibilities). The output is the source of the specified lexical analyzer in the C programming language. A specification in *lex* is mainly a sequence of pattern-action statement as in *awk*. Actions are pieces of C code to be inserted in the lexical analyzer. At run time, these pieces of code execute the action corresponding to the associated pattern, when found. The following line is a typical statement of a *lex* program:

```
[A-Za-z]+([A-Za-z0-9])*  { yyval = Install(); return(ID);}
```

The pattern specifies identifiers, that is, strings of characters starting with one letter and containing only letters and digits. This action leads the generated lexical analyzer to store the identifier and to return the string type "ID" to the calling parser. It is another instance of the regular expression-matching problem. The question of constructing **pattern-matching automata** is an important component having a practical application in the *lex* software.

Texts such as books or programs are likely to be changed during elaboration. Even after their completion they often support periodic upgrades. These questions are related to **text comparisons.** Sometimes we also wish to find a string, and do not completely remember it. The search has to be performed with an entirely non-specified pattern. This is an instance of the **approximate pattern matching.** Keeping track of all consecutive versions of a text may not be helpful because the text can be very long and changes may be hard to find. The reasonable way to control the process is to have an easy access to differences between the various versions. There is no universal notion as to what the differences are, or conversely, what the similarities are, between two texts. However, it can be agreed that the intersection of the two texts is the longest common subtext of both. In our book this is called the **longest common subsequence** problem, so that the differences between the two texts are the respective complements of the common part. The UNIX command *diff* builds on this notion. An option of the command *diff* produces a sequence of *ed* instructions to transform one text into the other. The similarity of texts can be measured as the minimal number of edit operations to transform one text into the other. The computation of such a measure is an instance of the **edit distance** problem.

1.2 Dictionaries

The search of words or patterns in static texts is quite a different question than the previous pattern-matching mechanism. Dictionaries, for example, are organized in order to speed up the access to entries. Another example of the same question is given by indexes. Technical books often contain an index of chosen terms that gives pointers to parts of the text related to words in the index. The algorithms involved in the creation of an index form a specific group. The use of dictionaries or lexicons is often related to natural language processing. Lexicons of programming languages are small, and their representation is not a difficult problem during the development of a compiler. To the contrary, English contains approximately 100,000 words, and even twice that if inflected forms are considered. In French, inflected forms produce more than 700,000 words. The representation of lexicons of this size makes the problem a bit more challenging.

A simple use of dictionaries is illustrated by spelling checkers. The UNIX command, *spell*, reports the words in its input that are not stored in the lexicon. This rough approach does not yield a pertinent checker, but, practically, it helps to find typing errors. The lexicon used by *spell* contains approximately 70,000 entries stored within less than 60 kilobytes of random-access memory. Quick access to lexicons is a necessary condition for producing good parsers. The data structure useful for such access is called an index. In our book indexes correspond to data structures representing all factors of a given (presumably long) text. We consider problems related to the construction of such structures: **suffix trees, directed acyclic word graphs, factor automata, suffix arrays**. The *PAT* tool developed at the NOED Center (Waterloo, Canada) is an implementation of one of these structures tailored to work on large texts. There are several applications that effectively require some understanding of phrases in natural languages, such as data retrieval systems, interactive software, and character recognition.

An image scanner is a kind of photocopier. It is used to give a digitized version of an image. When the image is a page of text, the natural output of the scanner must be in a digital form available to a text editor. The transformation of a digitized image of a text into a usual computer representation of the text is realized by an Optical Character Recognition (OCR). Scanning a text with an OCR can be 50 times faster than retyping the text on a keyboard. Thus, OCR softwares are likely to become more common. But they still suffer from a high degree of imprecision. The average rate of error in the recognition of characters is approximately one percent. Even if this may happen to be rather small, this means that scanning a book produces approximately one error per line. This is compared with the usually very high quality of texts checked by specialists. Technical improvements on the hardware can help eliminate certain kinds of errors occurring on scanned texts in printed forms. But this cannot alleviate the problem associated with recognizing texts in printed forms. Reduction of the number of errors can thus only be achieved by considering the context of the characters, which assumes some understanding of the structure of the text. Image processing is related to the problem of **two-dimensional pattern matching.** Another related problem is the data structure for all subimages, which is discussed in this book in the context of the **dictionary of basic factors.**

The theoretical approach to the representation of *lexicons* is either by means of trees or finite state automata. It appears that both approaches are equally efficient. This shows the practical importance of the **automata theoretic approach** to text problems. At LITP (Paris) and IGM (Marne-la-Vallée) we have shown that the use of automata to represent lexicons is particularly efficient. Experiments have been done on a 700,000 word lexicon of LADL (Paris). The representation supports direct access to any word of the lexicon and takes only 300 kilobytes of random-access memory.

1.3   Data compression

One of the basic problems in storing a large amount of textual information is the **text compression** problem. Text compression means reducing the representation of a text. It is assumed that the original text can be recovered from its compressed from. No loss of information is allowed. Text compression is related to the **Huffman coding problem** and the **factorization problem.** This kind of compression contrast with other kinds of compression techniques applied to sounds or images, in which approximation is acceptable. Availability of large mass storage does not decrease the interest for compressing data. Indeed, users always take advantage of extra available space to store more data or new kinds of data. Moreover, the question remains important for storing data on secondary storage devices. Examples of implementations of dictionaries reported above show that **data compression** is important in several domains related to natural language analysis. Text compression is also useful for telecommunications. It actually reduces the time to transmit documents via telephone network, for example. The success of Facsimile is perhaps to be credited to compression techniques.

General compression methods often adapt themselves to the data. This phenomenon is central in achieving high compression ratios. However, it appears, in practice, that methods tailored for specific data lead to the best results. We have experimented with this fact on data sent by geostationary satellites. The data have been compressed to seven percent of their original size without any loss of information.

The compression is very successful if there are redundancies and regularities in the information message. The analysis of data is related to the problem of **detecting regularities in texts**. Efficient algorithms are particularly useful to expertise the data.

1.4 Applications of text algorithms in genetics

Molecules of nucleic acids carry a large segment of information about the fundamental determinants of life, and, in particular, about the reproduction of cells. There are two types of nucleic acids known as desoxyribonucleic acid (DNA) and ribonucleic acid (RNA). DNA is usually found as double-stranded molecules. In vivo, the molecule is folded up like a ball of string. The skeleton of a DNA molecule is a sequence on the four-letter alphabet of nucleotides: adenine (A), guanine (G), cytosine (C), and thymine (T). RNA molecules are usually single-stranded molecules composed of ribonucleotides: A, G, C, and uracil (U).

Processus of "transcription" and "translation" lead to the production of proteins, which also have a string composed of 20 amino acids as a primary structure. In a first approach all these molecules can be viewed as texts. The discovery twenty years ago of powerful sequencing techniques has led to a rapid accumulation of sequence data. From the collection of sequences up to their analysis many algorithms on texts are implied. Moreover, only fast algorithms are often feasible because of the huge amount of data involved.

Collecting sequences can be accomplished through audioradiography gels. The automatic transcription of these gels into sequences is a typical **two dimensional pattern-matching** problem in two dimensions. The reconstruction of a whole sequence from small segments, used for instance in the shotgun sequencing method, is another example of a problem that occurs during this step. This problem is called the **shortest common superstring problem**: construction of the shortest text containing several given smaller texts.

Once a new sequence is obtained, the first important question to ask is whether it resembles any other sequence already stored in data banks. Before adding a new molecular sequence into an existing data base one needs to know whether or not the sequence is already present. The comparison of several sequences is usually realized by writing one over another. The result is know as an alignment of the set of nucleotides. Alignment of two sequences is the **edit distance problem**: compute the minimal number of edit operations to transform one string into another. It is realized by algorithms based on dynamic programming techniques similar to the one used by the UNIX command *diff*.

The problem of the **longest common subsequence** is a variation of the alignment of sequences. A tool, called *agrep*, developed at the University of Arizona, is devoted to these questions, related to **approximate string matching**.

Further questions about molecular sequences are related to their analysis. The aim is to discover the functions of all parts of the sequence. For example, DNA sequences contain important regions (coding sequences) for the production of proteins inside the cell. However, no good answer is presently known for finding all coding sequences of a DNA sequence. Another question about sequences is the reconstruction of their three-dimensional structure. It seems that a part of the information resides in the sequence itself. This is because, during the folding process of DNA, for example, nucleotides match pairwise (A with T, and C with G). This produces approximate palindromic symmetries (as TTAGCGGCTAA). Involved in all these questions are **approximate searches** for specific patterns, for **repetitions**, for **palindromes**, or other **regularities**.

1.5   Efficiency of algorithms

Efficient algorithms can be classified according to what is meant by efficiency. There exist different notions of efficiency depending on the complexity measure involved. Several such measures are discussed in this book: sequential time, memory space, parallel time, and number of processors.

This book deals with "feasible" problems. We can define them as problems having efficient algorithms, or as solvable in time bounded by a small-degree polynomial. In the case of sequential computations we are interested in lowering the degree of the polynomial corresponding to time complexity. The most efficient algorithms usually solve a problem in linear-time complexity. We are also interested in space complexity. Optimal space complexity often means a constant number of (small integer) registers in addition to input data. Therefore, we say that an algorithm is time-space optimal if it works simultaneously in linear time and in constant extra space. These are the most advanced sequential algorithms, and also the most interesting, both from a practical and theoretical point of view.

In the case of parallel computations we are generally interested in the parallel time $T(n)$ as well as in the number of processors $P(n)$ required for the executions of the parallel algorithm on data of size n. The total number of elementary operations performed by the parallel algorithm is not greater than the product $T(n)P(n)$.

Efficient parallel algorithms are those that operate in no more than poly-logarithmic (a polynomial of logs of input size) time with a polynomial number of processors. The class of problems solvable by such algorithms is denoted by NC and hence we call the related algorithms NC-algorithms. An NC-algorithm is optimal if the total number of operations $T(n)P(n)$ is linear. Another possible definition is that this number is essentially the same as the time complexity of the best known sequential algorithm solving the given problem. However, we adopt the first option here because algorithms on strings usually have a time complexity which is at least linear.

Precisely evaluating the complexity of an algorithm according to some measure is often difficult, and, moreover, it is unlikely to be of much use. The "big O" notation clarifies what the important terms of a complexity expression are. It estimates the asymptotic order of the complexity of an algorithm and helps compare algorithms between each others. Recall that if f and g are two functions from and to integers, then we say that $f = O(g)$ if $f(n) < C.g(n)$ when $n > N$, for some constants C and N. We write $f = \Theta(g)$ when the functions f and g are of the same order, which means that both equalities $f = O(g)$ and $g = O(f)$ hold.

Comparing functions through their asymptotic orders leads to these kinds of inequalities: $O(n^{0.7}) < O(n) < O(n \log n)$, or $O(n^{\log n}) < O(\log n^n) < O(n!)$.

Within sequential models of machines one can distinguish further types of computations: off-line, on-line and real-time. These computations are also related to efficiency. It is understood that real-time computations are more efficient than general on-line, and that on-line computations are more efficient than off-line. Each algorithm is an off-line algorithm: "off-line" conceptually means that the whole input data can be put into the memory before the actual computation starts. We are not interested then in the intermediate results computed by the algorithm, but only in the final result (though this final result can be a sequence or a vector). The time complexity is measured by the total time from the moment the computation starts (with all input data previously memorized) up to the final termination. In contrast, an on-line algorithm is like a sequential transducer. The portions of the input data are "swallowed" by the algorithm step after step, and after each step an intermediate result is expected (related to the input data read so far). It then reads the next portion of the input, and so on. In on-line algorithms the input can be treated as an infinite stream of data, consequently we are not interested mainly in the termination of the algorithm for all such data. The main interest for us is the total time $T(n)$ for which we have to wait to get the n-th first outputs. The time $T(n)$ is measured starting at the beginning of the whole computation (activation of the transducer). Suppose that the input data is a sequence and that after reading the n-th symbol we want to print "1" if the text read to this moment contains a given pattern as a suffix, otherwise we print "0". Hence we have two streams of data: the stream of input symbols and an output stream of answers "1" or "0". The main feature of the on-line algorithm is that it has to give an output value before reading the next input symbol. The real-time computations are those on-line algorithms that are in a certain sense optimal; the elapsing time between reading two consecutive input symbols (the time spent for computing only the last output value) should be bounded by a constant. Most linear on-line algorithms are in fact real-time algorithms.

We are primarily interested in off-line computations in which the worst-case running time is linear, but on-line and real-time computations, as well as average complexities are also discussed in this book.

1.6   Some notation and formal definitions

Let A be an input *alphabet*–a finite set of symbols. Elements of A are called the *letters*, the *characters*, or the *symbols*.   Typical examples of alphabets are: the set of all ordinary letters, the set of binary digits, or the set of 256 8-bit ASCII symbols. Texts (also called words or strings) over A are finite sequences of elements of A. The length (size) of a text is the number of its elements (with repetitions). Therefore, the length of *aba* is 3. The length of a word x is denoted by $|x|$. The input data for our problems will be words, and the size n of the input problem will usually be the length of the input word. In some situations, n will denote the maximum length or the total length of several words if the input of the problem consists of several words.

The i-th element of the word x is denoted by $x[i]$ and i is its *position* on x. We denote by $x[i..j]$ the factor $x[i]x[i+1]...x[j]$ of x. If $i > j$, by convention, the word $x[i..j]$ is the empty word (the sequence of length zero), which is denoted by ε.

We say that the word x of length m is a factor (also called a subword) of the word y if $x = y[i+1..i+n]$ for some integer i. We also say that x *occurs in y at position i*, or that the position i is a *match* for x in y.

We define the notion of *subsequence* (sometimes called a subword). The word x is a subsequence of y if x can be obtained from y by removing zero or more (not necessarily adjacent) letters from it. Likewise, x is a subsequence of y if $x = y[i_1]y[i_2]...y[i_m]$, where $i_1, i_2, ..., i_m$ is an increasing sequence of indices on y.

Next we define formally the basic problem covered in this book. We often consider two texts *pat* (the pattern) and *text* of respective lengths m and n.

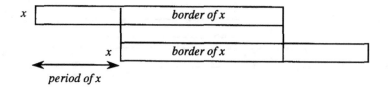

Figure 1.1: Duality between periods and borders of texts.

String matching (the basic problem). Given texts *pat* and *text*, verify if *pat* occurs in *text*. This is a decision problem: the output is a Boolean value. It is usually assumed that $m \leq n$. Therefore, the size of the problem is n. A slightly advanced version entails searching for all occurrences of *pat* in *text*, that is, computing the set of positions of *pat* in *text*. Let us denote this set by *MATCH(pat, text)*. In most cases an algorithm computing *MATCH(pat, text)* is a trivial modification of a decision algorithm, this is the reason why we sometimes present only decision algorithms for string matching.

Instead of just one pattern, one can consider a finite set of patterns and ask if a given text contains a pattern from the set. The size of the problem is now the total length of all patterns plus the length of the text.

1.7 Some simple combinatorics of strings

The main theoretical tools in string-matching algorithms are related to mathematical properties of periodicities in strings. We define the notion of period of a word, which is central in almost all strings matching algorithms. A *period* of a word x is an integer p, $0 < p \leq |x|$, such that

$$x[i] = x[i + p]$$

for all $i \in \{1, \ldots, |x| - p\}$. When there is no ambiguity, we also say that the word $x[1..p]$ is a period of x. This is the usual definition of a period for a function defined on integers, as x can be viewed. Note that the length of a word is always a period of it, so that any word has at least one period. We denote by *period(x)* the smallest period of x. We additionally say that x is *periodic* if $period(x) \leq |x|/2$.

The notion of *border* of a text is a dual notion to that of period, see Figure 1.1. A border of x is any word that is simultaneously a prefix and a suffix of x. Observe that x and the empty string ε are borders of x.

Let us denote by *Border(x)* the longest nontrivial border (not the whole

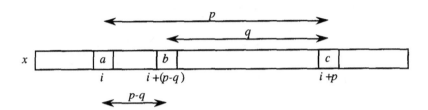

Figure 1.2: Quantity $p - q$ is also a period because letters a and b are both equal to letter c.

word) of x. Note that

$$(|x| - |Border(x)|, |x| - |Border^2(x)|, \ldots, |x| - |Border^k(x)|)$$

is the sequence of all periods of x in increasing order (k is the smallest integer for which $Border^k(x)$ is the empty word).

Example. The periods of *aabaaabaa* (of length 9) are 4, 7, 8 and 9. Its corresponding proper borders are *aabaa, aa, a, ε*.

Periodicity Lemma

Let x be a non-empty word and p be an integer such that $0 < p \leq |x|$. Then each of the following conditions equally defines p as a period of x:

1. x is a factor of some word y^k with $|y| = p$ and $k > 0$,
2. x may be written $(uv)^k$ with $|uv| = p$, v a non-empty word, and $k > 0$,
3. for some words y, z and w, $x = yw = wz$ and $|y| = |z| = p$.

Lemma 1.1 [Periodicity Lemma] *Let p and q be two periods of the word x. If $p + q < |x|$, then $\gcd(p, q)$ is also a period of x.*

Proof. The conclusion trivially holds if $p = q$. Assume now that $p > q$. First we show that the condition $p + q < |x|$ implies that $p - q$ is a period of x. Let $x = x[1]x[2] \ldots x[n]$ ($x[i]$'s are letters). Given $x[i]$ the i-th letter of x, the condition implies that either $i - q \geq 1$ or $i + p \leq n$. In the first case, q and p being periods of x, $x[i] = x[i-q] = x[i-q+p]$. In the second case, for the same reason, $x[i] = x[i+p] = x[i+p-q]$. Thus $p - q$ is a period of x. This situation is shown in Figure 1.2. The rest of the proof, left to the reader, is by induction on the integer $\max(p, q)$, after noting that $\gcd(p, q)$ equals $\gcd(p - q, q)$. □

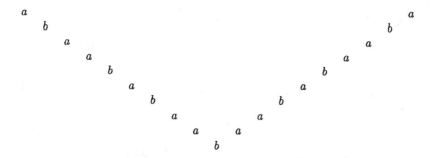

Figure 1.3: After cutting off its last two letters, Fib_8 is a symmetric word, a *palindrome*. This is not accidental.

There is a stronger version of the periodicity lemma for which we omit the proof.

Lemma 1.2 [Strong Periodicity Lemma] *If p and q are two periods of a word x such that $p + q - \gcd(p, q) \leq |x|$, then $\gcd(p, q)$ is also a period of x.*

An interesting family: Fibonacci words

Fibonacci words form an interesting family of words (from the point of view of periodicities). In sone sense, the inequality that appears in Strong Periodicity Lemma is optimal. The example supporting this claim is given by the *Fibonacci words* with the last two letters deleted.

Let Fib_n be the n-th Fibonacci word ($n \geq 0$). It is defined by

$$Fib_0 = \varepsilon,\ Fib_1 = b,\ Fib_2 = a,\text{ and } Fib_n = Fib_{n-1}Fib_{n-2},\text{ for } n > 2.$$

Fibonacci words satisfy a large number of interesting properties related to periods and repetitions. Note that Fibonacci words (except the first two words of the sequence) are prefixes of their successors. Indeed, there is an even stronger property: the square of any Fibonacci word of high enough rank is a prefix of its succeeding Fibonacci words. Among other properties of Fibonacci words, it must be noted that they have no factor in the form u^4 (u non empty word) and they are almost symmetric, see Figure 1.3. Therefore, Fibonacci words contain a large number of periodicities, but none with an exponent higher than 3.

The lengths of Fibonacci words are the well-known Fibonacci numbers, $f_0 = 0$, $f_1 = 1$, $f_2 = 1$, $f_3 = 2$, $f_4 = 3$, ... The first Fibonacci words of the sequence $(Fib_n, n > 2)$ are

$$Fib_3 = ab, \qquad\qquad\qquad\qquad\qquad\qquad |Fib_3| = 2,$$
$$Fib_4 = aba, \qquad\qquad\qquad\qquad\qquad\qquad |Fib_4| = 3,$$
$$Fib_5 = abaab, \qquad\qquad\qquad\qquad\qquad\quad |Fib_5| = 5,$$
$$Fib_6 = abaababa, \qquad\qquad\qquad\qquad\quad\;\; |Fib_6| = 8,$$
$$Fib_7 = abaababaabaab, \qquad\qquad\qquad\quad |Fib_7| = 13,$$
$$Fib_8 = abaababaabaababaababa, \qquad\quad |Fib_8| = 21,$$
$$Fib_9 = abaababaabaababaababaabaababaabaab, \quad |Fib_9| = 34.$$

1.8 Some other interesting strings

Fibonacci words of rank greater than 1 can be treated as prefixes of a single infinite Fibonacci string Fib_∞. Similarly we can define the words of Thue-Morse $T(n)$ as prefixes of a single infinite word T_∞. Assume we count positions on this word starting from 0. Denote by $g(k)$ the number of "1" in the binary representation of the number k. Then

$$T_\infty(k) = \begin{cases} a & \text{if } g(k) \text{ is even,} \\ b & \text{otherwise.} \end{cases}$$

The Thue-Morse words T_n are the prefixes of T_∞ of length 2^n. We list several of them below.

$$T_1 = ab,$$
$$T_2 = abba,$$
$$T_3 = abbabaab,$$
$$T_4 = abbabaabbaababba,$$
$$T_9 = abbabaabbaababbabaababbaabbabaab.$$

These words have the remarkable property of being *overlap-free*, which means that there is no nonempty word x that occurs in them at two positions which distance is smaller than $|x|$. However these words are mostly known for the following *square-free* property: they contain no nonempty word in the form xx (nor, indeed, in the form $axaxa$, $a \in A$).

Let us define the following invertible encoding:

$$\beta(a) = a, \quad \beta(b) = ab, \quad \text{and } \beta(c) = abb.$$

Lemma 1.3 *For each integer n the word $\beta^{-1}(T_n)$ is square free.*

The lemma says in particular that there are infinitely many "square-free" words. Let T'_∞ be the word over the alphabet $\{0, 1, 2\}$ which symbols are the number of occurrences of letter "b" between two consecutive occurrences of letter "a" in T_∞. Then such an infinite word is also "square-free". We have

$$T'_\infty = 2\ 1\ 0\ 2\ 0\ 1\ 2\ \ldots$$

Other interesting words are sequences of moves in the *Hanoi towers* game. There are six possible moves depending from which stack to which other stack an disk is moved. If we have n disks then the optimal sequence consists of $2^n - 1$ moves and forms a word H_n. The interesting property of these words is that all of them are "square-free".

Yet another family of words that has a strange relation to numbers $g(k)$ is given by the binary words P_n, where P_n is the n-th row of the Pascal triangle modulo 2. In other words:

$$P_n(i) = \binom{n}{i} \mod 2.$$

We list below some of these words.

$$
\begin{aligned}
P_0 &= & 1 \\
P_1 &= & 1 \quad 1 \\
P_2 &= & 1 \quad 0 \quad 1 \\
P_3 &= & 1 \quad 1 \quad 1 \quad 1 \\
P_4 &= & 1 \quad 0 \quad 0 \quad 0 \quad 1 \\
P_5 &= & 1 \quad 1 \quad 0 \quad 0 \quad 1 \quad 1
\end{aligned}
$$

The word P_n has the following remarkable property: the number of "1" in P_n equals $2^{g(n)}$.

Let us consider the infinite string W which symbols are digits and which results from concatenating all consecutive natural numbers written in decimal. Hence,

$$W = 0123456789101112131415161718192021222324252627282930313 2\ldots$$

Denote by W_n the prefix of W of size n. For a word x, let us denote by $occ_n(x)$ the number of occurrences of x in W_n. The words W_n have the following interesting property: for every two nonempty words x and y of a same length

$$\lim_{n\to\infty} \frac{occ_n(x)}{occ_n(y)} = 1.$$

This means, in a certain sense, that the sequence W is quite *random*.

An interesting property of strings is how many factors of a given length k they contain. Assume the alphabet is $\{a, b\}$. For a given k we have 2^k different words of length k. A natural question is:

what is the minimal length $\gamma(k)$ of a word containing each subword of length k.

Obviously $\gamma(k) \geq 2^k + k - 1$, since any shorter word has less than 2^k factors. It happens that $\gamma(k) = 2^k + k - 1$. The corresponding words are called de Bruijn words. In these strings each word of length k occurs exactly once. For a given k there are exponentially many de Bruijn words. For example for $k = 1$ we can take ab, for $k = 2$ we take $aabba$ or $abaab$ and for $k = 3$ we can take de Bruijn word $aaababbbaa$.

There is an interesting relation of de Bruijn words to Euler cycles in special graphs G_k. The nodes of G_k are all words of length $k - 1$ and for any word $x = a_1 a_2 \ldots a_{k-1}$ of length $k - 1$ we have two directed edges

$$a_1 a_2 \ldots a_{k-1} \xrightarrow{a} a_2 \ldots a_{k-1} \cdot a, \qquad a_1 a_2 \ldots a_{k-1} \xrightarrow{b} a_2 \ldots a_{k-1} \cdot b$$

The graph has a directed Euler cycle (containing each edge exactly once). Let $a_1 a_2 \ldots a_N$ be the sequence of labels of edges in a Euler cycle. Observe that $N = 2^k$. As de Bruijn word we can take the word:

$$a_1 a_2 \ldots a_N a_1 a_2 \ldots a_{k-1}.$$

1.9   Cyclic shifts and primitive words

A cyclic shift of x is any word vu, when x can be written in the form uv. Let us consider how many different cyclic shifts a word can have.

Example. Consider the cyclic shifts of the word $abaaaba$ of length 7. There are exactly 7 different cyclic shifts of $abaaaba$, the 8-th shift goes back to the initial word.

a	b	a	a	a	b	a		
	b	a	a	a	b	a	a	
		a	a	a	b	a	a	b
			a	a	b	a	a	b	a
				a	b	a	a	b	a	a
					b	a	a	b	a	a	a
						a	a	b	a	a	a	b
							a	b	a	a	a	b	a

A word w is a said to be *primitive* if it is not of the form $w = v^k$, for a natural number $k \geq 2$. As a consequence of the *periodicity lemma* we show the following fact.

Lemma 1.4 *Assume the word x is primitive. Then x has exactly $|x|$ different* *cyclic shifts. In other words:*

$$|\{vu \ : \ u \ and \ v \ words \ such \ that \ x = uv \ and \ u \ne \varepsilon\}| \ = \ |x|.$$

Proof. Assume x of length p has two cyclic shifts that are equal. Hence $x = uv = u'v'$, and $vu = v'u'$, where $u \ne u'$.

Assume without loss of generality that $|u'| < |u|$. Then $u = u'\alpha$, $v' = \alpha \cdot v$ and $vu'\alpha = \alpha vu'$. Hence the text $\alpha \cdot v \cdot u' \cdot \alpha$ has borders $\alpha \cdot v \cdot u'$ and α. Consequently, the text $\alpha \cdot v \cdot u' \cdot \alpha$ has two periods of size $r = |\alpha|$ and $p = |vu'\alpha|$. At the same time $r + p = |\alpha \cdot v \cdot u' \cdot \alpha|$.

The periodicity lemma implies that the text has period $\gcd(r, p)$. Since $r < p$ this shows that p is divisible by the length of the smaller period. This implies that x is a power of a smaller word, which contradicts the assumption. Consequently x cannot have two identical cyclic shifts. □

We show a simple number-theoretic application of primitive words and cyclic shifts. In 1640 the great French number theorist Pierre de Fermat stated the following theorem.

Theorem 1.1 [Fermat's Simple Theorem] *If p is a prime number and n is any natural number then p divides $n^p - n$.*

Proof. Define the equivalence relation $\equiv$ on words by $x \equiv y$ if x is a cyclic shift of y. A word is said to be unary if it is in a form a^p, for a letter a. Take the set S of all non-unary words of length p over the alphabet $\{1, 2, \dots, n\}$. All these words are primitive since their length is a prime number and they are non-unary. According to Lemma 1.4 each equivalence class has exactly p elements. The cardinality of S is $n^p - n$ and S can be partitioned into disjoint subsets of the same cardinality p. Hence the cardinality of S is divisible by p, consequently $n^p - n$ also is. This completes the proof. □

Bibliographic notes

Complementary notions, problems and algorithms in stringology may be found in the books by Crochemore and Rytter [CR 94], by Stephen [St 94], by Gusfield [Gu 97], by Crochemore, Hancart and Lecroq [CHL 01], and in the collective book edited by Apostolico and Galil [AG 97].

Chapter 2

Basic string searching algorithms

The string-matching problem is the most studied problem in algorithmics on words, and there are many algorithms for solving this problem efficiently. We assume that the pattern *pat* is of length m, and that the text *text* has length n, both are given as *read-only* arrays. Two basic string-matching algorithms are Knuth-Morris-Pratt (*KMP*) algorithm and Boyer-Moore (*BM*) algorithm. Each of them consists of two phases:

pattern-preprocessing phase: computing certain tables related to the pattern: *Bord*, *Strong_Bord*, *BM_Shift*,

searching phase: finding the first one or all occurrences of *pat* in *text*.

In this chapter we present the searching phases of both algorithms together with searching phases of their variations. The preprocessing phases are more technical; they are included in the next chapter. We begin with a scheme of a brute-force algorithm that uses quadratic time. Such a naive algorithm is, in fact, an origin of *KMP* and *BM* algorithms. The informal scheme of such a naive algorithm is:

$(*)$ **for** $i := 0$ **to** $n - m$ **do**
             check if $pat = text[i + 1..i + m]$.

2.1   Knuth-Morris-Pratt algorithm

The actual implementation of (∗) differs with respect to how we implement the *checking* operation: scanning the pattern from the left or scanning the pattern from the right (or otherwise). We then get two brute-force algorithms. Both algorithms have quadratic worst-case complexity. In this section we discuss the first of them (left-to-right scanning of the pattern).

To shorten the presentation of some algorithms we assume that the statement **return**(x) outputs the value of x and stops the whole algorithm.

If $pat = a^{n/2}b$ and $text = a^{n-1}b$ then in the algorithm *brute-force1* a quadratic number of symbol comparisons takes place. However, the average complexity is not so high.

    Algorithm *brute-force1*;
      $i := 0$;
    while $i \leq n - m$ **do begin**
        $j := 0$; { left-to-right scan of *pat* }
        while $j < m$ **and** $pat[j + 1] = text[i + j + 1]$ **do**
          $j := j + 1$;
        if $j = m$ **then return**(true);
        { $inv1(i, j)$ } $i := i+1$; { length of shift = 1 }
    end;
    return(false) { there was no return earlier }

Our first linear-time algorithm is a natural, improved version of the naive algorithm *brute-force1*. We present a constructive proof of the following.

Theorem 2.1 *The string-matching problem can be solved in $O(|text| + |pat|)$ time using $O(|pat|)$ space. The constants involved in "O" notation are independent of the size of the alphabet.*

Remark. We are disappointed with this theorem on one point. The size of additional memory is rather large though linear in the size of the pattern. We show later that a constant number of registers suffices to achieve linear-time complexity (again the size of the alphabet does not intervene).

Let us look more closely at the algorithm *brute-force1* and at its main invariant

$$inv1(i, j): \; pat[1..j] = text[i + 1..i + j] \text{ and } (j = m \text{ or }$$
$$pat[j + 1] \neq text[i + j + 1]).$$

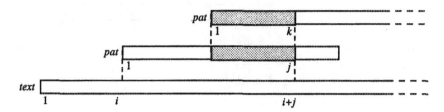

Figure 2.1: Shifting the pattern to the next safe position.

In fact, we first use the slightly weaker invariant

$$inv1'(i,j): \ pat[1..j] = text[i+1..i+j].$$

The invariant essentially says that the value of j gives us a lot of information about the last part of the text scanned up to this point.

Morris-Pratt algorithm

Using the invariant $inv1'(i,j)$, we are able to make longer shifts of the pattern. Present shifts in algorithm *brute-force1* always have length 1. Let s denote (the length of) a *safe shift*, where "safe shift s" means that, based on the invariant, we know that there is no occurrence of the pattern at positions between i and $i + s$, but there may be one at position $i + s$.

Assume $j > 0$, let $k = j - s$, and suppose an occurrence of the pattern starts at position $i + s$. Then, $pat[1..k]$ and $pat[1..j]$ are suffixes of the same text $text[1..i+j]$, see Figure 2.1. Hence, the following condition is implied by $inv1'$:

$$cond(j,k): \ pat[1..k] \text{ is a proper suffix of } pat[1..j].$$

Therefore the shift is safe if k is the largest position satisfying $cond(j,k)$. Denote this position by $Bord[j]$. Hence the smallest positive safe shift is

$$MP\_Shift[j] = j - Bord[j].$$

The function (table) *Bord* is called a *failure function* because it helps us at the time of a failure (mismatch). It is the crucial function. It is stored in a table with the same name. We also call this table the *table of borders*. The failure function allows us to compute the length of the smallest safe shift, which is $s = j - Bord[j]$.

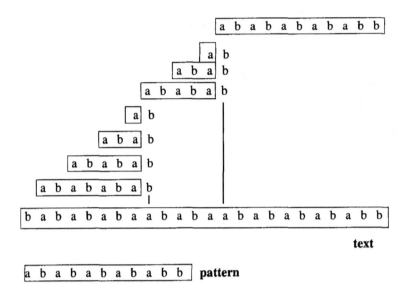

Figure 2.2: The history of algorithm *MP* on an example of text and pattern.

Note that *Bord*[j] is precisely the length of the largest proper border of *pat*[1..j], *Border*(*pat*[1..j]), as defined in Chapter 1. The longest proper border of a word x is the longest (non trivial) overlap when we try to match x with itself.

Example. For *pat* = *ababababb* we have:

$$Bord = [0, 0, 1, 2, 3, 4, 5, 6, 7, 8, 0].$$

The same pattern is used in Figure 2.2.

In the case $j = 0$, that is, when *pat*[1..j] is the empty word, we have a special situation. Since, in this situation, the length of the shift must be 1, we then define *Bord*[0] = −1. An improved version of algorithm *brute-force1* is the Morris-Pratt algorithm (*MP*) below.

Lemma 2.1 *The time complexity of the algorithm MP is linear in the length of the text. The maximal number of character comparisons executed is $2n - m$.*

Proof. Let $T(n)$ be the maximal number of symbol comparisons "*pat*[j + 1] = *text*[$i+j+1$]?" executed by the algorithm *MP*. There are at most $n-m+1$ unsuccessful comparisons (at most one for any given i). Consider the sum

$i + j$. Its maximal value is n and minimal value is 0. Each time a successful comparison is made the value of $i + j$ increases by one unit. This value, observed at the time of comparing symbols, never decreases. Hence, there are at most n successful comparisons. If the first comparison is successful then we have no unsuccessful comparison for position $i = 0$. We conclude that: $T(n) < n + n - m = 2n - m$. For $pat = ab$ and $text = aaaa \dots a$ we have $T(n) = 2n - m$ □

Algorithm *MP*; { algorithm of Morris and Pratt }
    $i := 0; \ j := 0;$
   while $i \leq n - m$ **do begin**
      while $j < m$ **and** $pat[j + 1] = text[i + j + 1]$ **do**
        $j = j + 1;$
      if $j = m$ **then return**(true);
      $i := i + MP\_Shift[j]; \quad j := \max(0, j - MP\_Shift[j]);$
   end;
   return(false)

Knuth-Morris-Pratt algorithm

We have not yet taken into account the full invariant *inv*1 of algorithm *brute-force1*, but only its weaker version *inv*1′. We have left the *mismatch* property apart. We now develop a new version of algorithm *MP* that incorporates the *mismatch* property:

$$pat[j + 1] \neq text[i + j + 1].$$

The resulting algorithm, called *KMP*, improves the number of comparisons performed on a given letter of the text. The clue for improvement is the following: assume that a mismatch in algorithm *MP* occurs on the letter $pat[j + 1]$ of the pattern. The next comparison is between the same letter of the text and $pat[k + 1]$ if $k = Bord[j]$. But if $pat[k + 1] = pat[j + 1]$, the same mismatch appears. Therefore, we can avoid considering the border of $pat[1 . . j]$ of length k in this situation.

For $m > j \geq 0$ consider a condition stronger than $cond(j, k)$ by a *one-comparison* information:

$strong\_cond(j, k)$: ($pat[1 . . k]$ is a proper suffix of $pat[1 . . j]$ and
                       $pat[k + 1] \neq pat[j + 1]$).

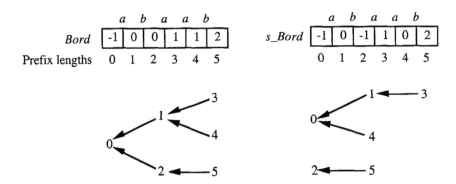

Figure 2.3: Functions *Bord* and *Strong_Bord* for pattern *abaab*.

We then define $Strong\_Bord[j]$ as k, where k is the smallest integer satisfying $strong\_cond(j, k)$, and as -1 otherwise. Moreover, we define $Strong\_Bord[m]$ as $Bord[m]$. We say that $Strong\_Bord[i]$ is the length of the longest *strong border* of $pat[1..j]$. Figure 2.3 illustrates the difference between functions *Bord* and *Strong_Bord* on pattern *abaab*.

Algorithm *KMP*; { algorithm of Knuth, Morris and Pratt }
    $i := 0; \ j := 0;$
    while $i \leq n - m$ **do begin**
        while $j < m$ **and** $pat[j + 1] = text[i + j + 1]$ **do**
            $j = j + 1;$
        if $j = m$ **then return**(true);
        $i := i + KMP\_Shift[j];$
        $j := \max(0, j - KMP\_Shift[j]);$
    end;
    return(false)

The algorithm *KMP* is the algorithm *MP* in which table *Bord* is replaced by table *Strong_Bord* and *MP_Shift* is replaced by:

$$KMP\_Shift[j] = j - Strong\_Bord[j].$$

The history of the algorithm is shown in Figure 2.4. The table *Strong_Bord* is more effective in the on-line version of algorithm *KMP* below (*on-line-KMP*). Assume that the text ends with the special end-marker *end-of-text*. Each time

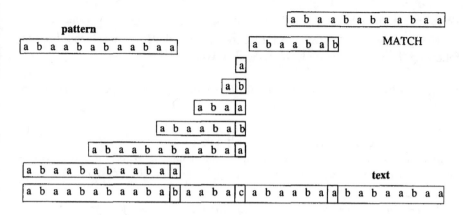

Figure 2.4: The history of algorithm *KMP* on example strings.

we process the current input symbol, we then output 1 if the part of the text read so far ends with the pattern *pat*; otherwise we output 0.

Denote by *delay*(*m*) the maximal time, measured as the number of statements

$$j := Strong\_Bord[j]$$

elapsed between two consecutive reads, for patterns of length *m*. By *delay'*(*m*), we denote the time corresponding to the use of *Bord* instead of *Strong_Bord*.

```
Algorithm on-line-KMP;
  { on-line linear version of KMP search }
    read(symbol); j := 0;
    while symbol ≠ end-of-text do begin
      while j < m and pat[j + 1] = symbol do begin
        j := j + 1; if j = m then write(1) else write(0);
        read(symbol);
      end;
      if Strong_Bord[j] = −1 then begin
        write(0); read(symbol); j := 0;
      end else
        j := Strong_Bord[j];
    end
```

The large gap between *delay*(*m*) and *delay'*(*m*) can be seen on the following example: *pat* = *aaaa*...*a* and *text* = $a^{m-1}b$. In this case, *delay*(*m*) = 1 while

$delay'(m)$ is the length of *pat*. The value of $delay(m)$ is generally small.

Using properties of text periodicities presented in the next chapter the following lemma can be deduced (see also [KMP 77]).

Lemma 2.2 *For KMP algorithm* $delay(m) = O(\log m)$, *and the bound is tight.*

Observe that for texts on binary alphabets $delay(m)$ is constant. This means that in this case, the algorithm is real-time. However, if patterns are over the alphabet $\{a, b\}$ and texts over $\{a, b, c\}$, then the delay can be logarithmic.

It is an interesting exercise to program transformations that modify the algorithm on-line-KMP to achieve a real-time computation independently of the size of the alphabet. This means that the time between two consecutive reads must be bounded by a constant. The crucial observation is that if we execute "$j := Strong\_Bord[j]$," then we know that for the next $j - Strong\_Bord[j]$ input symbols the output value will be 0 ("no match"). This allows for dispersal of output actions between input actions (reading symbol) in such a way that the time between consecutive writes-reads is bounded by a constant. To do so, we can maintain up to m last symbols of the text *text* in a table. We leave details to the reader. It is interesting to observe that the real-time condition can be achieved by using any of the tables *Bord*, *Strong_Bord*. In this way, we sketched the proof of the following result (which becomes much more difficult if the model of computation is a Turing Machine).

Theorem 2.2 *There is a real-time algorithm for string matching on a Random Access Machine.*

2.2 Boyer-Moore algorithm and its variations

In this section we describe yet another basic approach to string matching. We can get another naive string-matching algorithm, similar to *brute-force1*, if the scan of the pattern is done from right to left. This algorithm has quadratic worst-case behavior, but (similarly to algorithm *brute-force1*) its average-time complexity is linear. In this section we discuss a derivative of *brute-force2*: the Boyer-Moore algorithm. The main feature of this algorithm is that it is efficient in the sense of worst-case (for most variants) as well as average-case complexity. For most texts and patterns the algorithm scans only a small part of the text because it performs "jumps" on the text. The algorithm *brute-force2* wastes information related to the invariant:

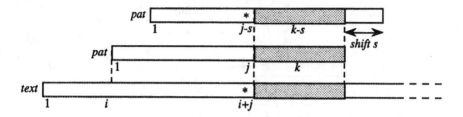

Figure 2.5: The case $s < j$: $s = BM\_Shift[j]$.

$inv2 : pat[i+1..m] = text[i+j+1..i+m]$ and $pat[i] \neq text[i+j]$.

```
Algorithm brute-force2;
    i := 0;
    while i < n − m do begin
        j := m; { right-to-left scan of pat }
        while j > 0 and pat[j] = text[i + j] do
            j := j − 1;
        if j = 0 then return true;
        { inv2(i, j) } i := i + 1; { length of shift = 1 }
    end;
    return false;
```

The information gathered by the algorithm is "stored" at the value of j. Suppose that we want to make better shifts using the invariant. A shift s is said to be *safe* if we are certain that between i and $i+s$ there is no starting position of the pattern in the text. Suppose that the pattern appears at position $i + s$ (see figure 2.5), where the case $s < j$ is presented). Then, the following conditions hold:

$cond1(j, s)$: for each k such that $j < k \leq m$, $s > k$ or $pat[k − s] = pat[k]$,
$cond2(j, s)$: if $s < j$ then $pat[j − s] \neq pat[j]$ { mismatch property }.

We define two kinds of shifts, each associated with a suffix of the pattern represented by a position $j < m$, and defined by its length:

$Weak\_BM\_Shift[j] = \min\{s > 0 : cond1(j, s) \text{ holds}\}$,
$BM\_Shift[j] = \min\{s > 0 : cond1(j, s) \text{ and } cond2(j, s) \text{ hold}\}$.

We also define

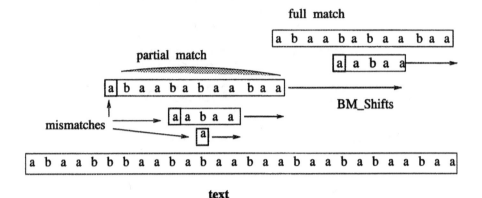

Figure 2.6: The history of algorithm BM on example strings.

$$Weak\_BM\_Shift[m] = BM\_Shift[m] = m - Bord[m] = period(pat).$$

BM algorithm is a version of *brute-force2* in which, in a mismatch situation, a shift of length $BM\_Shift[j]$ is executed instead of only a one-position shift (see Figure 2.6).

Algorithm BM;
{ improved version of *brute-force2* }
   $i := 0$;
   while $i \leq n - m$ **do begin**
      $j := m$;
      while $j > 0$ **and** $pat[j] = text[i + j]$ **do**
         $j := j - 1$;
      if $j = 0$ **then return** true;
      { $inv2(i, j)$ } $i := i + BM\_Shift[j]$;
   end;
   return false;

Compare a run of this algorithm with a run of the similar algorithm that uses $Weak\_BM\_Shift$ instead of BM_Shift. Take as an example the strings

$$pat = cababababa \quad \text{and} \quad text = aaaaaaaaaababababa.$$

For strings of similar structure the algorithm BM makes $O(n)$ comparisons while its weaker version (when the table $Weak\_BM\_Shift$ is used instead of

BM_Shift), makes $\Omega(n^2)$ comparisons. This shows that, contrary to the behavior of MP and KMP algorithms, the utilization of the mismatch property is here crucial to achieve linear-time complexity. In the next chapter we prove the following nontrivial fact.

Theorem 2.3 . *Algorithm BM makes $O(n)$ comparisons to find the first occurrence of a pattern in a text of length n.*

Boyer and Moore introduced also another "heuristic" useful in increasing lengths of shifts. Suppose that we have a situation, where $symb = text[i+j]$ (for $j > 0$), and $symb$ does not occur at all in the pattern. Then, in the mismatch situation, we can make a shift of length $s = j$. For example, if $pat = a^{100}$ and $t = (a^{99}b)^{10}$, then we can always shift 100 positions, and eventually make only 10 symbol comparisons. For the same input words, algorithm BM makes 901 comparisons.

If we take $pat = ba^{m-1}$ and $text = a^{2m-1}$, the heuristic used alone (without using table BM_Shift) leads to a quadratic number of comparisons. Let $LAST(symb)$ be the last position of an occurrence of symbol $symb$ in pat. If there is no occurrence, $LAST(symb)$ is set to zero. Then, we can define a new shift, replacing instruction "$i := i + BM\_Shift[j]$" of BM algorithm by "$i := i + max(BM\_Shift[j], j - LAST(text[i+j]))$."

The shift of length $j - LAST(text[i+j])$ is called an *occurrence shift*. In practice, it may improve the time of the search for some inputs, though theoretically it is not entirely analyzed. If the alphabet is binary, and more generally for small alphabets, the occurrence heuristic has little effect.

BM algorithm is a simple as well as a very efficient algorithm. Its beauty relies upon its simplicity, and this is somehow partially lost when we optimize this algorithm. On the other hand the efficiency can be improved. We describe two algorithms in which inspection of the pattern starts from the right end of the pattern (the main feature of BM).

If we wish to find all occurrences of pat in $text$ with algorithm BM (trivially modified to report all occurrences), then the complexity can become quadratic. The simplest example is given by a text and a pattern over a one-letter alphabet. Observe a characteristic feature of this example: high periodicity of the pattern. Let p be the period of the pattern. If we discover an occurrence of pat at some position in $text$, then the next shift must naturally be equal to p. Afterward, we have only to check the last p symbols of pat. If they match with the text, then we can report a complete match without inspecting all other $m - p$ symbols of pat. This simple idea is embodied in the algorithm below. The variable named $memory$ "remembers" the number of symbols that we do not have to inspect ($memory = 0$ or $memory = m - p$). In fact, it remembers the prefix of the pattern that matches the text at the current position. This technique is called *prefix memorization*. The correctness of the following algorithm is straightforward. The period of *pat* can be precomputed once for all searches for *pat*.

Algorithm *BMG*;
{ *BM* algorithm with prefix memorization }
$\quad p = period(pat) = BM\_Shift[0]$ }
$i := 0$; *memory* := 0;
while $i \leq n - m$ **do begin**
$\quad j := m$
$\quad$ **while** $j > memory$ **and** $pat[j] = text[i + j]$ **do**
$\quad\quad j := j - 1$;
$\quad$ **if** $j = memory$ **then begin**
$\quad\quad$ write(i); *memory* := $m - p$;
$\quad$ **end else** *memory* := 0;
$\quad$ { $inv2(i, j)$ } $i := i + BM\_Shift[j]$;
end;

Theorem 2.4 *Algorithm BMG makes $O(n)$ comparisons.*

Proof. We shall prove in Chapter 3 (by a complicated argument) that the number of comparisons to find the first occurrence is $O(n' + m)$, where n' is the position of the first occurrence. This implies that the complexity to find all occurrences is $O(n + r.m)$, where r is the number of occurrences of *pat* in *text*. This is because between any two consecutive occurrences of *pat* in *text*, *BMG* does not make more comparisons than the original *BM* algorithm. Hence, if $p \geq 2m/2$, since $r < n/p$, $n + r.m$ is $O(n)$. We have yet to consider the case $p < m/2$. In this case, we can group occurrences of the pattern into chains of positions distant only by p (for two consecutive positions in a group).
Within each such chain every text symbol is inspected at most only once. The gaps between chains are larger than $m/2$, and, inside each such gap, *BMG* does not work slower than *BM* algorithm. An argument similar to that used in the case of large periods can now be applied.   □

BM algorithm is particularly fast for alphabets that are large relatively to the length of the pattern, because shifts are likely to be long. For small alphabets, the average number of symbol comparisons is linear. We design next an algorithm making $O(n\frac{\log m}{m})$ comparisons on the average. Hence, if m is of the same order as n, the algorithm makes only $O(\log n)$ comparisons. It is essentially based on the same strategy as *BM* algorithm, and can be treated as another variation of it.

For simplicity, we assume that the alphabet has only two elements, and that each symbol of the text is chosen independently with the same probability. Let $r = 2\lceil \log m \rceil$.

Theorem 2.5 *The algorithm fast-on-average runs in $O(n \log m/m)$ expected time and (simultaneously) in $O(n)$ worst-case time if the pattern is preprocessed. The preprocessing of the pattern takes $O(m)$ time.*

Proof. A preprocessing phase is needed to efficiently check if $text[i - r . . i]$ is a factor of *pat* in $O(r)$ time. Any of the data structures developed in Chapters 4, 5 and 6 (a suffix tree or a DAWG) can be used. Assume that *text* is a random string. There are $2^{r+1} > m^2$ possible segments of *text*, and less than m factors of *pat* of length r. Hence, the probability that the suffix of length r of *text* is a factor of *pat* is not greater than $1/m$. The expected time spent in each of the subintervals $[1 . . m]$, $[m - r . . 2.m - r - 1]$, ... is $O(m.1/m + r) = O(r)$ (these are consecutive subintervals of size m that have overlap of size r). There are $O(n/m)$ such intervals. Thus, the total expected time of the algorithm is of order $(r + 1)n/m = n(\log m + 1)/m$. □

Algorithm *fast-on-average*;
    $i := m$;
  while $i \le n$ **do begin**
    if $text[i - r . . i]$ is a factor of *pat* **then**
        compute all occurrences of *pat* which starting
        positions are in $[i - m . . i - r]$ applying *KMP* algorithm
    else { pattern does not start in $[i - m . . i - r]$ }
    $i := i + m - r$;
  end

Bibliographic notes

MP algorithm is from Morris and Pratt [MP 70]. The fundamental algorithm considered in this chapter (*KMP* algorithm) has been designed by Knuth, Morris, and Pratt [KMP 77]. Our exposition is slightly different than in this paper. A criterion that says whether an on-line algorithm can be transformed into a real-time algorithm has been shown by Galil in [Ga 81]. The principle applies to *MP* and *KMP* algorithms.

Algorithm *BM* is originally from Boyer and Moore [BM 77]. The variant *BMG* is from Galil [Ga 79]. Another interesting variation of Boyer-Moore algorithm is the algorithm *Turbo-BM*, see [C-R 92]. The additional memory is only increased by two integer variables storing the last shift and the size of the last matched part. The algorithm makes at most $2n$ letter comparisons. Apostolico and Giancarlo [AG 86] designed a variant using an additional memory of size proportional to the pattern length and that makes no more than $2n/3$ symbol comparisons (see [CL 97]).

Chapter 3

Preprocessing for basic searchings

In this chapter we discuss the preprocessing phases for the algorithms Knuth-Morris-Pratt, Boyer-Moore and their variations. Some combinatorics of words is needed for the analysis of Boyer-Moore algorithm. This analysis is rather sophisticated and can be omitted in the first reading.

3.1 Preprocessing patterns for *MP* and *KMP* algorithms

The preprocessing for the algorithms *MP* and *KMP* consists in the computation of the tables of borders and strong borders. We start with the computation of the table *Bord*, and our aim is to derive a linear-time algorithm. We present in this subsection two solutions. The first approach is to use algorithm *MP* to compute *Bord*. This, at first instance, can appear contradictory because *Bord* is needed inside the algorithm. However, we compute *Bord* in parts.

Whenever a value $Bord[j]$ is needed in the computation of $Bord[i]$ for $i > j$ then $Bord[j]$ is already computed. Suppose that $text = pat$ (or indeed $text = pat[2..m]$. We apply algorithm *MP* starting with $i = 1$ (for $i = 0$ nothing interesting happens) and continue with $i = 2, 3, \ldots, m - 1$. Then $Bord[r] = j > 0$ whenever $i + j = r$ is a successful comparison for the first time. If $Bord[r] > 0$ then such a comparison will take place. Assume that initially $Bord[i] = -1$ for all $j > 0$. Our interest here is only a side effect of the

33

algorithm *MP* (computation of the border table). The complexity of algorithm *compute-Borders1* is linear. The argument is the same as for algorithm *MP*.

```
procedure compute-Borders1;
{ a version of algorithm MP with text = pat }
    i := 1; j := 0;
    while i < m + 1 do begin
        while i + j < m and pat[j + 1] = pat[i + j + 1] do begin
            j := j + 1;
            if Bord[i + j] = −1 then Bord[i + j] := j;
        end;
        i := i + j − Bord[j]; j := max(0, Bord[j]);
    end
```

Next, we present the most classical linear-time algorithm computing table *Bord*.

```
procedure compute-Borders2;
{ computes the failure table Bord for pat, second version }
    Bord[0] := −1; t := −1;
    for j := 1 to m do begin
        while t ≥ 0 and pat[t + 1] ≠ pat[j] do t := Bord[t];
        t := t + 1; Bord[j] := t;
    end
```

The history of the algorithm is illustrated on an example string in Figure 3.1.

Lemma 3.1 *The maximum number of character comparisons executed by algorithm compute-Borders2 is $2.m - 3$.*

Proof. The complexity can be analyzed using a so-called "store principle." Interpret t as the number of items in a store. Note that when $t < 0$, no comparison is done, and t becomes null. Therefore, we can consider that the store is initially empty. For each j running from 2 to m, we add at most one item (at statement $t := t + 1$). However, whenever we execute the statement "$t := Bord[t]$," the value of t strictly decreases, which can be interpreted as deleting a nonzero number of items from the store. The total number of items inserted does not exceed $m - 2$. Hence, the total number of executions of

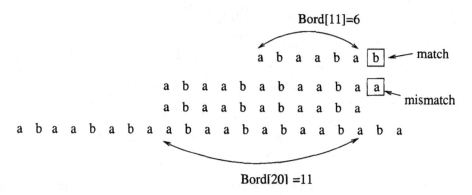

Bord[11]=6

a b a a b a [b] ⟵ match

a b a a b a b a a b a [a]
                                      mismatch
a b a a b a b a a b a

a b a a b a b a a b a a b a b a a b a b a

Bord[20] =11

Figure 3.1: Computation of $Bord[20] = Bord[Bord[19]]+1 = 7$ using procedure compute_borders2. The iteration for $j = 20$ starts with $t = Bord[19] = 11$.

statement "$t := Bord[t]$" for unsuccessful comparisons "$pat[t + 1] \neq pat[j]$" does not exceed $m - 2$. For each j running from 2 to m, there is at most one successful comparison. The total number of successful comparisons then does not exceed $m - 1$. Hence, the total number of comparisons does not exceed $2.m - 3$.   □

The computation of strong borders of prefixes of the pattern pat relies on the following observation. Let $t = Bord[j]$. Then, $Strong\_Bord[j] = t$ if $pat[t + 1] \neq pat[j + 1]$. Otherwise, the value of $Strong\_Bord[j]$ is the same as the value of $Strong\_Bord[t]$ because $pat[t + 1] = pat[j + 1]$. Note that the strong border of pat itself is its border, as if pat were followed by a marker.

```
procedure compute-strong-borders;
   { computes table Strong_Bord for pattern pat }
   Strong_Bord[0] := −1; t := −1;
   for j := 1 to m do begin { t equals Bord[j − 1] }
       while t ≥ 0 and pat[t + 1] ≠ pat[j] do t := Strong_Bord[t];
       t := t + 1;
       if j = m or pat[t + 1] ≠ pat[j + 1] then Strong_Bord[j] := t
       else Strong_Bord[j] := Strong_Bord[t];
   end
```

Example. Consider the pattern aba^{m-2}. Strong borders computed by the

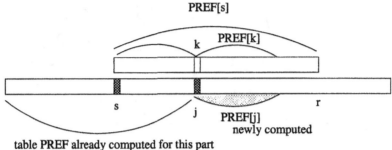

Figure 3.2: The typical situation when executing the procedure *Compute-Prefixes*.

procedure *compute-strong-borders* are given by the following table *Strong_Bord*:

$Strong\_Bord[0] = -1$, $Strong\_Bord[1] = 0$, $Strong\_Bord[2] = -1$, and $Strong\_Bord[j] = 1$, for $3 \leq j \leq m$.

This is a worst case for which exactly $3m - 5$ symbol comparisons are performed by the algorithm.

3.2 Table of prefixes

We introduce another useful table, denoted by *PREF*, related to the border table and *BM_Shift* table. It is defined by

$$PREF[i] = \max\{j : pat[i..i+j-1] \text{ is a prefix of } pat \}.$$

In the first algorithm we compute *PREF* scanning the pattern left-to-right. Assume we scan the j-th ($j > 1$) position and the following invariant is preserved (see Figure 3.2):

the values of $PREF[t]$ for $t < j$ are already computed, and
$s < j$ is a position such that $s + PREF[s] - 1$ is maximum.

We add a special end-marker at position $m+1$ on *pat* to simplify the description of the algorithm. We use an auxiliary function *Naive-Scan(p, q)*, such that

$$Naive\text{-}Scan(p, q) = \max\{k \geq 1 \text{ such that } pat[p..p+k-1] = pat[q..q+k-1]\}.$$

If there is no such $k > 0$ then *Naive-Scan(p, q)* $= 0$. Obviously the time complexity of *Naive-Scan(p, q)*, measured as number of comparisons, is at most $k + 1$, where k is the value returned by the function.

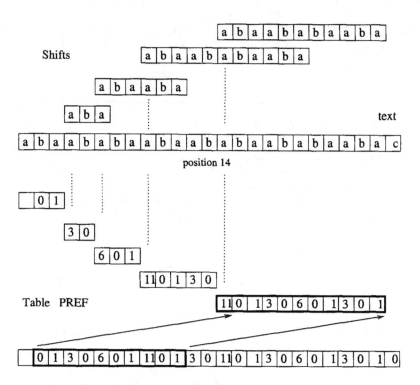

Figure 3.3: The behavior of the procedure *Compute-Prefixes* on an example string.

```
function Naive-Scan(p, q)
    result := 0;
    while (p ≤ n) and (q ≤ n) do begin
        if (t[p] ≠ t[q]) then break;
        p := p+1; q := q+1; result := result+1;
    end;
    return result;
```

Figure 3.3 illustrates the behavior of algorithm *Compute-Prefixes*. We have $PREF[14] = 11$, and $s = 14$, for the positions 15, 16, 17, ... , 25; for each such position j the corresponding $PREF[j]$ value is copied from the initial segment of the $PREF$ table as $PREF[j - 14 + 1]$.

For example $PREF[17] = PREF[17 - 14 + 1] = PREF[4] = 3$. This means that we do not need any new comparison for some values which are just duplicated.

```
procedure Compute-Prefixes;
    PREF[1] := 0; s := 1;
    for j := 2 to m do begin
        k := j - s + 1; r := s + PREF[s] - 1;
        if r < j then begin
            PREF[j] := Naive-Scan(j, 1); if PREF[j] > 0 then s := j;
        end else if PREF[k] + k < PREF[s] then
            PREF[j] := PREF[k]
        else begin
            x := Naive-Scan(r + 1, r - j + 2);
            PREF[j] := r - j + 1 + x; s := j;
        end
    end;
    PREF[1] := n;
```

Relation between tables of prefixes and of borders

Using the precomputed table *PREF* we can easily construct the tables of borders and strong borders in the procedure *compute-Borders3*.

```
procedure compute-Borders3;
    for k := 0 to m do Strong_Bord[k] := -1;
    for j := m down to 1 do begin
        i := j + PREF[j] - 1; Strong_Bord[i] := PREF[j];
    end;
    Bord[1] := 0; Bord[m] := Strong_Bord[m];
    for j := m - 1 down to 2 do
        Bord[i] := min{Bord[i + 1] - 1, Strong_Bord[i]};
```

A reverse computation is also possible: compute *PREF* knowing the table *Bord*. It is based on the following observation:

$$\text{if } Bord[i] = j \text{ then } pat[i - j + 1 .. i] \text{ is a prefix of } pat.$$

This leads to the following (almost correct) algorithm. Though useful, this algorithm is not entirely correct. In the case of one-letter alphabets only one entry of table *PREF* will be accurately computed. But its incorrectness is quite "weak." If $PREF[k] > 0$ after the execution of the algorithm then it is accurately computed. Moreover, *PREF* is computed for "essential" entries. The entry i is essential iff $PREF[i] > 0$ after applying this algorithm. These entries partition interval $[1..m]$ into subintervals for which further computation is relatively simple.

The computation of the table for nonessential entries is executed as follows: traverse the interval left-to-right and update $PREF[i]$ for each entry i. Take the first (to the left of i, including i) essential entry k with $PREF[k] \geq i-k+1$, then set

$$PREF[i] = \min\{PREF[i - k + 1], PREF[k] - (i - k)\}.$$

If there is no such essential entry, then $PREF[i] = 0$.

---

procedure *Alternative-Compute-PREF*;
{ correct for essential entries }
    for $i := 1$ **to** n **do** $PREF[i] := 0$;
    for $i := 1$ **to** n **do begin**
        $j := Bord[i]$; $PREF[i - j + 1] := \max(PREF[i - j + 1], j)$;
    end

Remark. Apply this for an example pattern over a one-letter alphabet to see how it works.

3.3   Preprocessing for Boyer-Moore algorithm

We show that the time complexity of computing the table *BM_Shift* is linear. First, using the table *PREF*, it is very easy to construct a linear-time algorithm for the computation of the table S of suffixes: $S[j]$ is the length of the largest suffix of the whole pattern ending at j. This table is used in the Apostolico-Giancarlo algorithm, and will be used in the precomputation of the table *BM_Shift*. It is easy to see that the computation of S is reducible to the computation of the table *PREF* for the reverse of the pattern.

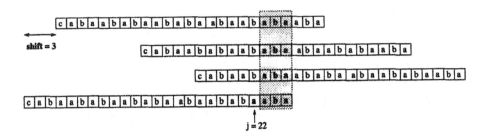

Figure 3.4: The case when $BM\_Shift[j] < j$. For $j = 22$, and an example string of size $m = 25$, we have $BM\_Shift[22] = \min\{m - k : j = m - S[k] = 22\}$. We have here $m - S[k] = 22$, so $S[k] = 3$. For $k = 9, 14, 22$, we have $S[9] = S[14] = S[22] = 3$, hence $BM\_Shift[22] = m - 22 = 25 - 22 = 3$.

procedure *compute-table-of-suffixes*(P);
  $P^R := reverse(P)$;
  compute table of prefixes $PREF^R$ for the word P^R;
  for each i **do** $S[i] := PREF^R[m - i + 1]$;

Observe that if $BM\_Shift[j] = m - k < j$, then $S[k] = m - j$. For example, for $j = 22$ and example pattern in Figure 3.4, $BM\_Shift[j] = 3$, and $S[25 - 3] = m - j = 3$. Using this observation and the table S, the shifts for BM algorithm are computed now as follows.

procedure *compute-BM-Shifts*;
  for $k := 1$ **to** $n - 1$ **do begin**
    $j := m - S[k]$; $BM\_Shift[j] := m - k$;
  end

The correct values of $BM\_Shift[j]$ are not computed here for the case when $BM\_Shift[j] > j$. In this case the mismatch property is ignored and the computation is reducible to computing borders of the whole pattern. We leave the consideration of this special (easy) case and the completion of the procedure *compute-BM-Shifts* to the reader.

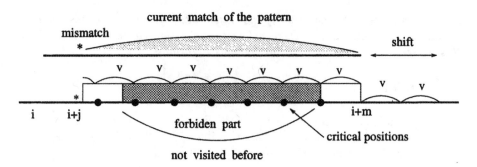

Figure 3.5: The part $text[i + j - 1 .. i + m]$ of the text is the current match; v denotes the shortest full period of the suffix of the pattern, v is a period of the current match. Shaded area is the "forbidden" part of the text.

3.4   * Analysis of Boyer-Moore algorithm

Now we analyze the Boyer-Moore algorithm.   The tight upper bound for the number of comparisons done by BM algorithm is approximately $3.n$. The proof of this is rather difficult but it yields a fairly simple proof of a $4.n$ bound. The fact that the bound is linear is completely non trivial, and surprising in view of the quadratic behavior of BM algorithm when modified to search for all occurrences of the pattern. The algorithm uses variable j to enlarge shifts, but afterwards "forgets" about the checked portion of the text. In fact, the same symbol in $text$ can be checked a logarithmic number of times.

If we replace BM_Shift by $Weak\_BM\_Shift$ then the time complexity becomes quadratic (a counterexample is given by text and patterns with the structure

$$pat = ca(ba)^k \text{ and } text = a^{2.k+2}(ba)^k.$$

Hence, one small piece of information (the "one bit" difference between invariants $inv2$, $inv2'$) considerably reduces the time complexity in the worst case. This contrasts with improved versions of algorithm $brute\text{-}force1$, where $inv1$ and $inv1'$ present similar complexities (see Chapter 2). The difference between the usefulness of information with one mismatched symbol gives evidence of the great importance of a (seemingly) technical difference in scanning the pattern left-to-right versus right-to-left.

Assume that, in a given non-terminating iteration, BM algorithm scans the part $text[i + j - 1 .. i + m]$ of the text and then makes the shift of length $s = BM\_Shift[j]$, where $j > 0$ and $s > (mj)/3$. By *current match*, we mean the scanned part of text without the mismatched letter (see Figure 3.5).

Lemma 3.2 *Let s be the value of the shift made in a given non-terminating iteration of BM algorithm. Then, at most $3.s$ positions of the text scanned at this iteration have been scanned in previous iterations.*

Proof. It is easier to prove a stronger claim:

($*$): positions in the segment $[i + j + k .. i + m - 2.k]$ on *text* are scanned for the first time during this iteration, where k is the size of the shortest full period v of $pat[m - s + 1 .. m]$.

In other words: only the first k and the last $2.k$ positions of the current match could have been scanned previously. Denote by v the shortest full period v of $pat[m - s + 1 .. m]$. The following property of the current match result from the definition of the shift:

(Basic property) v is a period of the current match, and v is a suffix of the pattern.

We introduce the notion of *critical position* in the current match. This is an internal position in this match in which the distance from the end of the match is a nonzero multiple of k, (see Figure 3.5). We say that a previous match ends at a position q in the text, if, in some previous iteration, the end of the pattern was positioned at q.

Claim 1. No previous match ends at a critical position of the current match.

Proof. (of the claim) The position $i + m$ is the end position of the current match. It is easy to see that if a critical point of the current match is the end of the match in a previous iteration i, then, in the iteration immediate after i, the end of the pattern is at position $i + m + shift$. Hence, the current match under consideration would not exist—a contradiction. This ends the proof of the claim. □

Claim 2. The length of the overlap of the current match and the previous match is smaller than k.

Proof. (of the claim) Recall that by a match we mean a scanned part of the text without the mismatch position. The period v is a suffix of the pattern. We already know, from Claim 1, that the end of the previous match cannot end at a critical position. Hence, if the overlap is at least k long, then v occurs inside the current match with the end position not placed at a critical position. The primitive word v then properly overlaps itself in a text in which the periodicity is v. But, this is impossible for primitive words (due to periodicity lemma). This proves the claim. □

Claim 3. Assume that a previous match ends at position q inside a forbidden area, and is completely contained in the current match. Then, there is no critical point (in the current match) to the right of q.

Proof. (of the claim) Suppose there is a critical position r to the right of q. It is then easy to see that $r - q$ is a good candidate for the shift in the *BM* algorithm. The algorithm takes the smallest such candidate as an actual shift. If the shift is smaller than $r - q$, we have a new position $q1 < r$. Then, we will have a sequence of previous matches with end positions $q1$, $q2$, $q3$, ... This sequence terminates in r, otherwise we would have an infinite increasing sequence of natural numbers smaller than r, which is impossible. This contradicts Claim 1, since we have a previous match ending at a critical position in the current match. This completes the proof of the claim. □

Proof of lemma. Now we are ready to show that $(*)$ holds. The proof is by contradiction. Assume that in some earlier iteration we scan the "forbidden" part of the text (shaded in Figure 3.5). Let q be the end position of the match in this iteration. Then q is not a critical position, and this match is contained completely in the current match (its overlap with the current match is shorter than k and q lies too far from the beginning of the current match). By the same argument, the rightmost critical position in the current match is to the right of q. Hence, we have found a previous match that is completely contained in the current match and in which the end position lies to the left of some critical position. This is impossible, however, due to Claim 3. This completes the proof of the lemma. □

Theorem 3.1 *The Boyer-Moore algorithm makes at most $4n$ symbol comparisons to find the first occurrence of the pattern (or to report no matches). The linear-time complexity of the algorithm does not depend on the size of the alphabet.*

Proof. The cost of each non-terminating iteration can be split into two parts:

1. the cost of scanning some positions of the text for the first time,
2. three times the length of the shift.

The total cost of all non-terminating iterations can be estimated by separately totaling all costs of type (1), this gives at most n, and all costs of type (2), which gives at most $3.(n - m)$. The cost of a terminating iteration is at most m. Hence, the total cost of all iterations is upper bounded by:

$$n + 3(n - m) + m \leq 4.n,$$

which completes the proof.   □

Bibliographic notes

The analysis of *KMP* algorithms is from Knuth, Morris, and Pratt [KMP 77].
The analysis of *BM* algorithm is from Cole [Co 91] [Co 77].

Chapter 4

On-line construction of suffix trees

We present here the first basic data structure representing the set $\mathcal{F}(text)$ of all factors (subwords) of a given word. Their importance derives from a multitude of applications. For simplicity we assume throughout this chapter that the alphabet A is of constant size (otherwise, the complexity of algorithms should be multiplied by $\log |A|$). Since $\mathcal{F}(text)$ is a set, the most typical problem related to such data structures is the *Membership Problem*:

$$\text{test if } x \in \mathcal{F}(text).$$

The data structure D representing the set $\mathcal{F}(text)$ is said to be *good* if:

(1) D has linear size,

(2) D can be constructed in linear time,

(3) D enables to test the membership problem in $O(|x|)$ time after preprocessing *text*.

4.1 Tries and their compact versions

Our approach to represent the set of factors of a text is graph theoretical. Let G be an acyclic rooted directed graph in which the edges are labeled with symbols or with words: $label(e)$ denotes the *label* of edge e. The *label* of a path π (denoted by $label(\pi)$) is the composition of labels of its consecutive edges. The edge-labeled graph G represents the set:

$Labels(G) = \{label(\pi) : \pi$ is a directed path in G starting at the root $\}$.

We also say that G represents the set of factors of *text* if $Labels(G) = \mathcal{F}(text)$.

The first *naive* approach to represent $\mathcal{F}(text)$ is to consider graphs that are trees in which edges are labeled by single symbols. These trees are called *subword tries*. Figure 4.1 shows the trie associated with the 6-th Fibonacci word $Fib_6 = abaababa$. In these trees, the links from a node to its children are labeled by letters. In the tree associated with *text*, a path down the tree spells a factor of *text*. All paths from the root to leafs spell suffixes of *text*. And all suffixes of *text* are labels of paths from the root. In general, these paths do not necessarily end in a leaf. The nodes correspond to subwords of the given text, each node can be identified with the word "spelled" by the path from the root to this node. In tries and suffix trees we distinguish nodes which correspond to suffixes of the given text, we call them *essential nodes*. Essential nodes are shaded in black in Figure 4.1.

Observation.   The tries are not "good" representations of $\mathcal{F}(text)$, because they can be too large. If $text = a^n b^n a^n b^n d$, then $Trie(text)$ has a quadratic number of nodes. We define a *chain* in a trie as a longest path consisting of non-essential nodes with outdegree one, except possibly the extremities of the chain. Two subtrees of a trie are isomorphic iff they have the same sets of paths leading from their roots to their essential nodes. We consider two kinds of succinct representations of the set $\mathcal{F}(text)$. They both result from compacting the tree $Trie(text)$. Two types of compaction can be applied, separately or simultaneously:

compacting chains, which produces the *suffix tree* of the text,

merging isomorphic subtrees (e.g., all leaves are to be identified), which leads to the *directed acyclic word graph* (DAWG) of the text, discussed in Chapter 6.

The suffix tree $\mathcal{ST}(text)$ is the compacted version of $Trie(text)$ when using the first method of compaction, see Figure 4.1 and Figure 4.2. Each chain π (path consisting of nodes of out-degree one) is compacted into a single edge e with $label(e) = [i, j]$, where $text[i..j] = label(\pi)$ (observe that a compact representation of labels is also used). Note a certain nondeterminism here, because there can be several possibilities of choosing i and j representing the same factor $text[i..j]$ of *text*. Any such choice is acceptable. In this context we identify the label $[i, j]$ with the word $text[i..j]$. The tree $\mathcal{ST}(text)$ is called the suffix tree of the word *text*.

For a node v of $\mathcal{ST}(text)$, let $val(v)$ be $label(\pi)$, where π is the path from the root to v. Whenever it is unambiguous we identify *nodes with their values*, and paths with their labels. Note that the suffix tree obtained by compacting

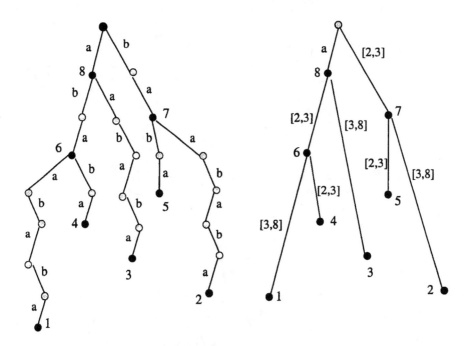

Figure 4.1: The tree *Trie(text)* and its compacted version, the suffix tree *ST(text)*, for the 6-th Fibonacci word: *abaababa*. The essential nodes are black. The numbers at these nodes indicate starting positions of the suffixes corresponding to the paths leading to these nodes. It is possible that the suffix tree contains some internal nodes with only one son since essential nodes are not deleted when compacting the trie.

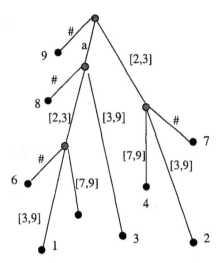

Figure 4.2: The suffix tree for the word of Figure 4.1 with end-marker: $\mathcal{ST}(abaababa\#)$. There are no internal nodes of out-degree one.

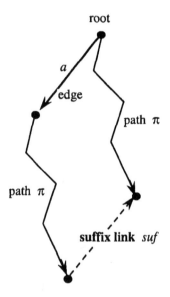

Figure 4.3: A suffix link *suf* points to the node representing the factor with the first letter removed, if such node exists.

chains has the following property: the labels of edges starting at a given node are words having different first letters. Therefore, the branching operation performed to visit the tree is reduced to comparisons on the first letters of the labels of outgoing edges. If we assume that no suffix of *text* is a proper prefix of another suffix (this is satisfied with a right end-marker), the leaves of *Trie(text)* are in one-to-one correspondence with the non-empty suffixes of *text*. The following fact about suffix trees is trivial, though a crucial one.

Lemma 4.1 *The size of the suffix tree $ST(text)$ is linear $(O(|text|))$.*

The crucial concept in efficient construction of tries and suffix trees is the table of *suffix links*: if x is the string corresponding to the node u then $suf[u]$ is a node which represents the string x with the first letter cut off, see Figure 4.3. If there is no such node then $suf[u] = $ nil. By convention, we define $suf[root] = root$.

4.2   Prelude to Ukkonen algorithm

We denote the prefix of length i of the text by p^i. We add a constraint on the suffix trie construction: not only do we want to build *Trie(p)*, but we also want to build on-line intermediate trees (see Figure 4.5)

$$Trie(p^1),\ Trie(p^2),\dots,\ Trie(p^{n-1}).$$

However, we do not keep all these intermediate suffix tries in memory, because overall it would take a quadratic time. Rather, we transform the current tree, and its successive values are exactly $Trie(p^1),\ Trie(p^2),\dots,\ Trie(p^n)$. Doing so we also require that the construction takes linear time (on fixed alphabets).

Let us examine closely how the sequence of uncompacted trees is constructed in Figure 4.5. The nodes corresponding to suffixes of the current text p^i are shaded. Let v_k be the suffix of length k of the current prefix p^i of the text. Identify v_k with its corresponding node in the tree. The nodes v_k are the *essential* nodes. In fact, additions to the tree (to make it grow up) are "essentially" created at such essential nodes. Consider the sequence $v_i, v_{i-1}, \dots, v_0$ of suffixes of p^i in decreasing order of their length. Compare such sequences in trees for the prefixes of the word *abaabb*, see Figure 4.5.

The basic property of the sequence of trees $Trie(p^i)$ is related to the way they grow. It is easily described with the sequence of essential nodes. Let $a_i = text[i]$ and let v_j be the first node in the sequence $v_{i-1}, v_{i-2}, \dots, v_0$ of essential nodes, such that $child(v_j, a_i)$ exists. Then, the tree $Trie(p^i)$ result from $Trie(p^{i-1})$ by adding a new outgoing edge labeled a_i, to each of the

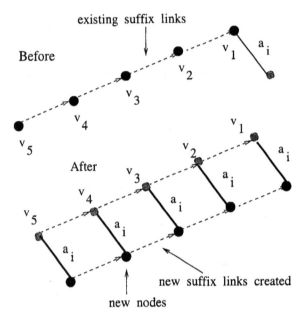

Figure 4.4: One iteration in the construction of *Trie(text)*. Thick edges are created at this step as well as new links for the corresponding new sons.

nodes $v_{i-1}, v_{i-2}, \ldots, v_{j-1}$ simultaneously creating a new son (see Figure 4.4). If there is no such node v_j, then a new outgoing edge labeled a_i is added to each of the nodes $v_{i-1}, v_{i-2}, \ldots, v_0$.

The sequence of essential nodes can be generated by iteratively taking suffix links from its first element, the leaf for which the value is the prefix read so far. The sequence of essential nodes is given by:

$$(v_i, v_{i-1}, \ldots, v_0) = (v_i, suf[v_i], suf^2[v_i], \ldots, suf^i[v_i]).$$

Using this observation we obtain the algorithm *on-line-trie*.

Theorem 4.1 *The algorithm on-line-trie builds the tree Trie(p) of suffixes of text on-line in time proportional to the size of Trie(p).*

Proof. The complexity results from the fact that the work performed in one iteration is proportional to the number of created edges. □

Algorithm *on-line-trie*;
    create the two-node tree $Trie(text[1])$ with suffix links;
    for $i := 2$ **to** n **do begin**
        $a_i := text[i]$;
        $v_{i-1} :=$ deepest leaf of $Trie(p^{i-1})$;
        $k := \min\{k : son(suf^k[v_{i-1}], a_i) \neq$ nil $\}$;
        create a_i-sons for $v_{i-1}, suf[v_{i-1}], \ldots, suf^{k-1}[v_{i-1}]$,
        and new suffix links for them (see Figure 4.4);
    end

4.3 Ukkonen algorithm

Ukkonen algorithm can be viewed as a "compacted version" of the algorithm *on-line-trie*. The basic point is that certain nodes of the trie do not exist explicitly in the suffix tree (after chain compaction). Indeed every node of the trie can be treated as an *implicit node* in the suffix tree. If this node corresponds to a node of the suffix tree then we say that such an implicit node is *real*. More formally: a pair (v, α) is an implicit node in T if v is a node of T and α is a (proper) prefix of the label of an edge from v to a son of it. The implicit node (v, α) is said to be a *"real"* node if α is the empty word.

Observation. The sequence of suffix trees produced by Ukkonen algorithm will differ slightly from our definition. We keep in the tree only the deepest internal essential node. Other essential nodes of out-degree one are not kept explicitly. However at the end of the algorithm we can add a special end-marker and this automatically will create all internal essential nodes. There is another important implementation detail. If the node v is a leaf, then there is no need to extend the edge coming from its father by a single symbol. If the label of the edge is a pair of positions (l, r), then, after updating it, it should be $(l, r + 1)$. We can omit such updates by setting r to $*$ for all leaves. This $*$-symbol is automatically understood as the last scanned position i of the pattern. Doing so reduces the amount of work involved at each iteration of the algorithm.

The i-th iteration executed in the previous algorithm can be adapted to work on a compact suffix tree $ST(p^{i-1})$. Each node of $Trie(p^{i-1})$ is treated as an implicit node. The new algorithm simulates the version *on-line-trie*.

If $v_i, v_{i-1}, \ldots, v_0$ is the sequence of essential nodes, and if we know that $v_i, v_{i-1}, \ldots, v_k$ are leaves, then we can skip processing them because this is done automatically by the $*$ trick. We thus start processing essential nodes from v_{k-1}. In the algorithm we call it v, the *working node*. This node is indicated in Figure 4.8. In other words the working node is the deepest internal essential node (corresponding to one suffix of the actual text). We do not maintain the set of essential nodes at lower levels. Another crucial concept is that of *implicit suffix* links, denoted by *imsuf*. If u is an implicit node then its suffix link can point to a non-real implicit node w, see Figure 4.6.

The computation of such a link is done by following the suffix link of the real-node father of u, then following down by the path having the same label as from $father(u)$ to u, see Figure 4.6. Creation of new edges is illustrated in Figure 4.9 and 4.10.

Theorem 4.2 *Ukkonen algorithm builds the compressed tree $ST(text)$ in an on-line manner. It works in linear time (on a fixed alphabet).*

Proof. The correctness follows from the correctness of the version working on an uncompacted tree. The new algorithm is just a simulation of it. To prove the $O(|text|)$ time bound, it is sufficient to prove that the total work is proportional to the size of $ST(text)$, which is linear. The work is proportional to the work performed by processing all working paths (paths of implicit suffix links needed to go from one working node to the working node of the next iteration). The cost of processing one working path is proportional to the decrease of distance between the working node and its father, plus some additional additive constant, see Figure 4.7. This distance is defined in terms of number of symbols from the implicit node (v, α) to its real father. If the working node is *real* itself then this distance is zero. On the other hand, the length of α is increased by at most one per iteration. Hence, the total increase of α is linear and, consequently, the total number of length reductions of α's is linear.   □

Algorithm *Ukkonen*;
  create the two-node tree $Trie(text[1])$ with suffix links;
  { v is the *working node* }
  $v := root$;
  for $i := 2$ **to** m **do begin**
    $a_i := text[i]$;
    if $son(v, a_i) \neq$ nil **then** $v := son(v, a_i)$
    else begin
      $k := \min\{k \ : \ son(imsuf^k[v], a_i) \neq \ \text{nil} \}$;
      create a_i-sons for v, $imsuf[v], \ldots, imsuf^{k-1}[v]$,
      and new $imsuf$ links for new internal nodes;
      $v := son(imsuf^{k-1}(v), a_i)$;
      { v is the deepest internal essential node of $Trie(p^{i-1})$ }
    end

Remark. If we want, at the end of the execution, to have a leaf corresponding to each suffix, then one extra stage of Ukkonen algorithm to manage an end-marker can do it, see Figure 4.10.

Bibliographic notes

The on-line algorithm that builds a compact suffix tree presented in the chapter has been discovered by Ukkonen [U 92]. The method is similar to the construction of suffix DAWG's presented in Section 6.2.

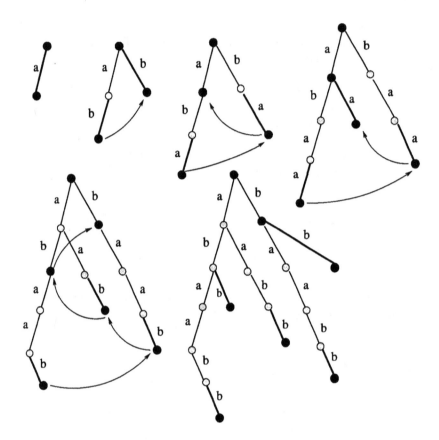

Figure 4.5: The history of the computation of the algorithm *on-line-trie*. The sequence of suffix tries $Trie(p^1)$, $Trie(p^2)$, ..., $Trie(p^6)$ for the text $p = abaabb$. Essential nodes are shaded. Edges created at the current step are thick. At each step i of the algorithm *on-line-trie* we follow a path of suffix links from the deepest node to the first node having an a_i-son.

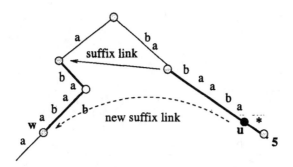

Figure 4.6: Computation of the implicit suffix link for an implicit node u, see also Figure 4.8.

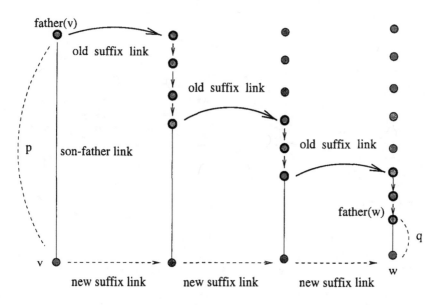

Figure 4.7: The working path in Ukkonen algorithm. Its cost can be charged to the decrease of the distance (measured in symbols) between the working node and its explicit father (the quantity $p - q$), plus an additive constant.

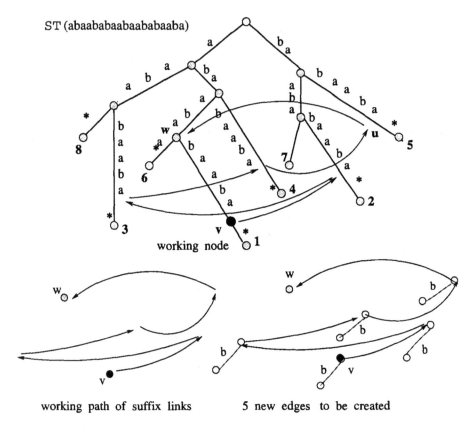

Figure 4.8: The tree during one stage of Ukkonen algorithm, the working node is shaded. The path of implicit suffix links is indicated by arrows. When letter b is added to the text, five new edges labelled by b are to be created.

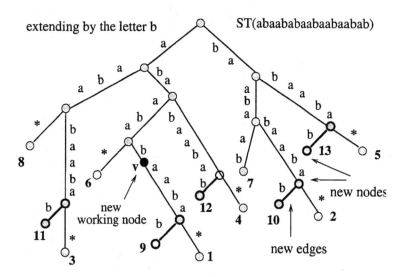

Figure 4.9: After extending the text by the letter b, several new edges, indicated by thick lines, are created.

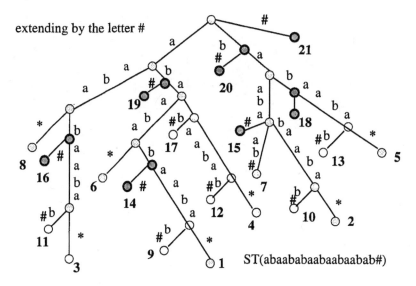

Figure 4.10: After extending the previous text by the end-marker # several new nodes (thick circles) are added to the tree. In this tree, each suffix corresponds to exactly one leaf.

Chapter 5

More on suffix trees

We present a few out of multitude of applications of suffix trees. In this chapter we sketch yet another renown algorithm for the suffix tree construction: McCreight algorithm. Assume the alphabet is of a constant size.

5.1 Several applications of suffix trees

There is a data structure slightly different from the suffix tree, known as the position tree. It is the tree of position identifiers. The *identifier* of position i on the text is the shortest prefix of $text[i..n]$ that does not occur elsewhere in *text*. Identifiers are well-defined when the last letter of *text* is a marker. Once we have the suffix tree of *text*, computing its position tree is fairly obvious (see Figure 5.1). Moreover, this shows that the construction works in linear time.

Theorem 5.1 *The position tree can be computed in linear time.*

One of the main applications of suffix trees is evident in the situation in which the text *text* is like a dictionary. In this situation, the suffix tree or the position tree acts as an index on the text. The index contains virtually all the factors of the text. With the data structure, the problem of locating a word w in the dictionary can be solved efficiently. But we can also perform other operations rapidly, such as computing the number of occurrences of w in *text*.

Theorem 5.2 *The suffix tree of text can be preprocessed in linear time so that, for a given word w, the following queries can be executed on-line in $O(|w|)$ time:*

- *find the first occurrence of w in text;*

59

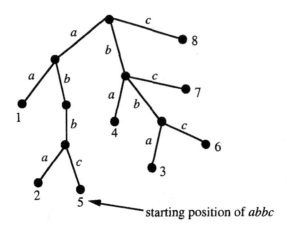

Figure 5.1: The position tree of $text = aabbabbc$.

- *find the last occurrence of w in text;*

- *compute the number of occurrences of w in text.*

We can list all occurrences of w in text in time $O(|w| + k)$, where k is the number of occurrences.

Proof. We can preprocess the tree, computing bottom-up, for each node, values *first*, *last* and *number*, corresponding respectively to the first position in the subtree, the last position in the subtree, and the number of positions in the subtree. Then, for a given word w, we can (top-down) retrieve this information in $O(|w|)$ time (see Figure 5.2). This gives the answer to the first three queries of the statement.

To list all occurrences, we first access the node corresponding to the word w in $O(|w|)$ time. Then, we traverse all leaves of the subtree to collect the list of positions of w in *text*. Let k be the number of leaves of the subtree. Since all internal nodes of the subtree have at least two sons, the total size of the subtree is less than $2.k$, and the traversal takes $O(k)$ time. This completes the proof.   □

The longest common factor problem is a natural example of a problem easily solvable in linear time using suffix trees, and very difficult to solve without any essential use of "good" representation of the set of factors. In fact, it has long been believed that no linear-time solution to the problem is possible, even if the alphabet is fixed.

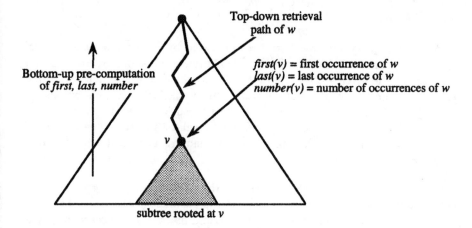

Figure 5.2: We can preprocess the tree to compute (bottom-up), for each node, its corresponding values *first*, *last*, and *number*. Then, for a given word w, we can retrieve this information in $O(|w|)$ time.

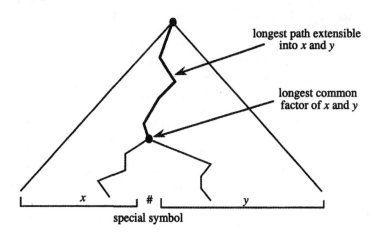

Figure 5.3: Finding longest common factors with suffix tree.

Theorem 5.3 *The longest common factor of k words can be found in linear time in the size of the problem, i.e., the total length of words (k is a constant, alphabet is fixed).*

Proof. The proof for case $k = 2$ is illustrated by Figure 5.3. The general case, with k fixed, is solved similarly. We compute the suffix tree of the text consisting of the given words separated by distinct markers. Then, exploring the tree bottom-up, we compute a vector informing for each node whether a leaf corresponding to a position inside the i-th subword is in the subtree of this node. The deepest node with positive information of this type for each i $(1 \leq i \leq k)$ corresponds to a longest common factor. The total time is linear. This completes the proof.   □

Suffix trees can be similarly applied to the problem of finding the longest repeated factor inside a given text. The solution is analogous to the solution of the longest common factor problem, considering that $k = 1$ and searching for a deepest internal node. Problems of this type (related to regularities in strings) is treated in more details in Chapter 8, where we show an algorithm having the running time $O(n \log n)$. This latter algorithm is simpler and does not use suffix trees or equivalent data structures. It also covers the two-dimensional case, where our "good" representations (considered in the present chapter) are not well suited.

Let $LCPref(i, j)$ denote the length of the longest common prefix starting at positions i and j in a given text of size n. In Chapter 14 on two-dimensional pattern matching we frequently use the following fact.

Theorem 5.4 *It is possible to preprocess a given text in linear time so that each query $LCPref(i, j)$ can be answered in constant time, for any positions i, j. A parallel preprocessing in $O(\log n)$ time with $O(n/\log n)$ processors of an EREW PRAM is also possible (on a fixed alphabet).*

Let $LCA(u, v)$ denote the *lowest common ancestor* of nodes u, v in a given tree T. The proof of Theorem 5.4 easily reduces to preprocessing the suffix tree that enables LCA queries in constant time. The value $LCPref(i, j)$ can be computed as the size of the string corresponding to the node $LCA(v_i, v_j)$, where v_i, v_j are leaves of the suffix tree corresponding to suffixes starting at positions i and j. In Chapter 7 we show (quite sofisticated) proof of the following thorem.

Theorem 5.5 *It is possible to preprocess a given tree T in linear time in such a way that each query $LCA(u, v)$ can be answered in constant time. A parallel preprocessing in $O(\log n)$ time with $O(n/\log n)$ processors of an EREW PRAM is also possible (on a fixed alphabet).*

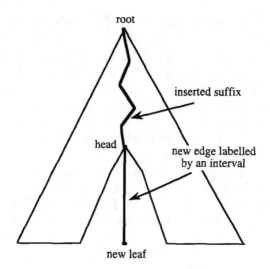

Figure 5.4: Insertion of a suffix into the tree.

We present also how to use suffix trees on the next problem: compute the number of distinct factors of *text* (cardinality of the set $\mathcal{F}(text)$).

Lemma 5.1 *We can compute the cardinality of the set $\mathcal{F}(text)$ in linear time.*

Proof. The weight of an edge in the suffix tree T is defined as the length of its label. Then, the required number is the sum of weights of all labels in the suffix tree.   □

5.2 McCreight algorithm

We present an optional construction of suffix trees. McCreight algorithm is an incremental algorithm. A tree is computed for a subset of consecutive suffixes consisting of all suffixes longer than some value. The next suffix is then inserted into the tree, and this continues until all (non-empty) suffixes are included in the tree.

Consider the structure of the path corresponding to a new suffix p inserted into the tree T. Such a path is indicated by the thick line in Figure 5.4. Denote by $insert(p, T)$ the tree obtained from T after insertion of the string p. The path corresponding to p in $insert(p, T)$ ends in the most recently created leaf of the tree. Denote the father of this leaf by *head*. It may be that the node *head* does not exist yet in the initial tree T (it is only an implicit node at the time of construction), and has to be created during the insert operation.

Let (w, α) be an implicit node of the tree T (w is a node of T, α is a word). The operation *break*(w, α) on the tree T is defined only if there is an edge outgoing node w in which the label δ has α as a prefix. Let β be such that $\delta = \alpha\beta$. The (side) effect of the operation *break*(w, α) is to break the corresponding edge; a new node is inserted at the breaking point, and the edge is split into two edges of respective labels α and β. The value of *break*(w, α) is the node created at the breaking point.

Let v be a node of the tree T, and let q be a subword of the input word *text* represented by a pair of integers l, r where $q = text[l..r]$. The basic function used in the suffix tree construction is the function *find*. The value *find*(v, q) is the last implicit node along the path starting at v and labeled by q. If this implicit node is not real, it is (w, α) for some non-empty α, and the function *find* converts it into the "real" node *break*(w, α).

Algorithm *Scheme of McCreight algorithm*;
   $T :=$ two-node tree with one edge labeled by $p_1 = text$;
   for $i := 2$ **to** n **do begin**
      { insert next suffix $p_i = text[i..n]$ }
      localize $head_i$ as $head(p_i, T)$,
         starting the search from $suf[father(head_{i-1})]$
         and using *fastfind* whenever possible;
      $T := insert(p_i, T)$;
   end

The important aspect of the algorithm is the use of two different implementations of the function *find*. The first one, called *fastfind*, deals with the situation when we know in advance that the searching path labeled by q is fully contained in some existing path starting at v. This knowledge allows us to find the searched node much faster using the compressed edges of the tree as shortcuts. If we are at a given node u, and if the next letter of the path is a, then we look for the edge outgoing u for which the label starts with a. Only one such edge exists due to the definition of suffix trees. This edge leads to another node u'. We jump in our searching path at a distance equal to the length of the label of edge (u, u'). The second implementation of *find* is the function *slowfind* that follows its path letter by letter. The application of *fastfind* is a main feature of McCreight algorithm, and plays a central part in its performance (together with links).

McCreight algorithm builds a sequence of compacted trees T_i in the order

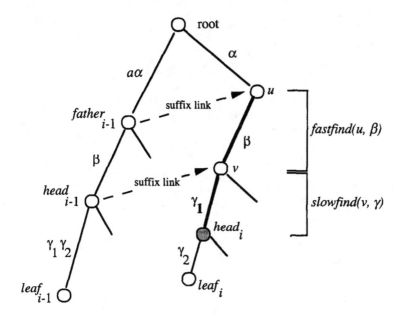

Figure 5.5: McCreight algorithm: the case where $v = head_i$ is a newly created node.

$i = 1, 2, \ldots, n$. The tree T_i contains the i-th longest suffixes of $text$. Note that T_n is the suffix tree $T(text)$, but that intermediate trees are not strictly suffix trees.

At a given stage of McCreight algorithm, we have $T \doteq T_{k-1}$ and we attempt to build T_k. The table of suffix links plays a crucial role in the reduction of the complexity. In the algorithm, the table suf is computed at a given stage for all nodes, except for leaves and maybe for the present head. The algorithm is based on the following two obvious properties:

1. $head_i$ is a descendant of the node $suf[head_{i-1}]$,

2. $suf[v]$ is a descendant of $suf[father(v)]$ for any v.

The basic work performed by McCreight algorithm involves localizing heads. If it is executed in a rough way (top-down search from the root), then the time is quadratic. The key to the improvement is the relation between $head_i$ and $head_{i-1}$, (Property 1). Hence, the search for the next head can start from some node deep in the tree, instead of from the root. This saves some work and the amortized complexity is linear. The behavior of McCreight algorithm is illustrated in Figures 5.5, 5.6, 5.7 and 5.8.

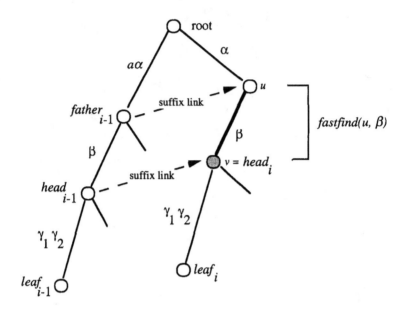

Figure 5.6: McCreight algorithm: the case where v is an existing node.

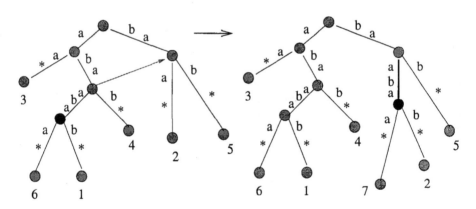

Figure 5.7: The tree of string $abaababaabaababaababa\$$, after inserting the first six suffixes, and the insertion of the 7-th suffix. The head in the left tree is $abaaba$ and, in the right one, it is $baaba$. The heads are indicated as black circles. In this case, $v = baaba$ is a newly created node, $\alpha = ba$, $\beta = aba$.

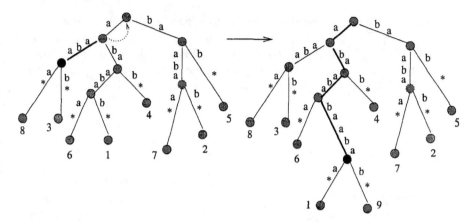

Figure 5.8: The tree of string *abaababaabaababaababa$*, after inserting the first eight suffixes, and the insertion of the 9-th suffix. In this case, $v = aaba$ is a newly created node, $\alpha = \varepsilon$, $\beta = aba$, and we perform a fastfind on the string *aba* and a slowfind on the string *ababaaba*.

Algorithm *McCreight*;
    $T :=$ two-node tree with one edge labeled by $p_1 = text$;
    for $i := 2$ **to** n **do begin**
        { insert next suffix $p_i = text[i..n]$ }
        let β be the label of the edge $(father[head_{i-1}], head_{i-1})$;
        let γ be the label of the edge $(head_{i-1}, leaf_{i-1})$;
        $u := suf[father[head_{i-1}]]$;
        $v := fastfind(u, \beta)$;
        $suf[head_{i-1}] := v$;
        if v has only one son **then**
            { v is a newly inserted node } $head_i := v$
        else $head_i := slowfind(v, \gamma)$;
        create a new leaf $leaf_i$; make $leaf_i$ a son of $head_i$;
        label the edge $(head_i, leaf_i)$ accordingly;
    end

Theorem 5.6 *McCreight algorithm runs in time $O(n)$ for constant alphabets.*

Proof. The total complexity of all executions of *fastfind* and *slowfind* is estimated separately. Let $father_i = father(head_i)$. The complexity of one run of *slowfind* at stage i is proportional to the difference $|father_i| - |father_{i-1}|$, plus some constant. Therefore, the total time complexity of all runs of *slowfind* is bounded by $\sum(|father_i| - |father_{i-1}|) + O(n)$. This is obviously linear. Similarly, the time complexity of one call to *fastfind* at stage i is proportional to the difference $|head_i| - |head_{i-1}|$, plus some constant. Therefore, the total complexity of all runs of *fastfind* is also linear.   □

Bibliographic notes

The two basic algorithms for suffix tree construction are from Weiner [We 73] and McCreight [McC 76]. Chen and Seiferas [CS 85] described the relation between DAWG's (Chapter 6) and suffix trees. An excellent survey on applications of suffix trees has been shown by Apostolico in [Ap 85]. As reported in [KMP 77], Knuth conjectured in 1970 that a linear-time computation of the longest common factor problem was impossible to achieve. Theorem 5.3 shows that it is indeed possible on a fixed alphabet. A direct computation of uncompacted position trees, without the use of suffix trees, is given is [AHU 74]. The algorithm is quadratic because uncompacted suffix trees can have quadratic size. The incremental algorithm for the suffix tree construction is from McCreight [McC 76].

Chapter 6

Subword graphs

The graph $DAWG(text)$, called the suffix DAWG of $text$, is obtained by identifying isomorphic subtrees of the uncompacted tree $Trie(text)$ representing $\mathcal{F}(text)$. We call this process *minimization*. In fact it corresponds to the minimization of finite deterministic automata. Observe that in the trie the nodes are classified as essential (corresponding to suffixes) and non-essential. This affects the minimization (see Figure 6.1). The reason to deal with DAWG's instead of minimal subword automata is that DAWG's are easier to construct on-line and their relation to suffix trees is more evident. Applications of suffix DAWG's are essentially the same as applications of suffix trees. We assume again in this chapter that the alphabet is of constant size.

6.1   Directed acyclic graph

Let x be a factor of $text$. We denote by $end\text{-}pos(x)$ (end positions) the set of all positions on $text$ where an occurrence of x ends. Let y be another factor of $text$. Then, the subtrees of $Trie(text)$ rooted at x and y (recall that we identify the nodes of $Trie(text)$ with their labels) are isomorphic if and only if $end\text{-}pos(x) = end\text{-}pos(y)$. In the graph $DAWG(text)$, paths having the same set of end positions lead to the same node. Hence, the nodes of G correspond to non-empty sets in the form $end\text{-}pos(x)$. The root of the DAWG corresponds to the whole set of positions $\{0, 1, 2, \ldots, n\}$ on the text. From a theoretical point of view, nodes of G can be identified with such sets (especially when analyzing the construction algorithm). But, from a practical point of view, the sets are never maintained explicitly. The *end-pos* sets, usually large, cannot directly name nodes of G, because the sum of the sizes of all such sets happens to be non linear.

The small size of DAWG's is due to the special structure of the family Φ of sets *end-pos*. We associate with each node v of the DAWG its value $val(v)$ equals to the longest word leading to it from the root. The nodes of the DAWG are, in fact, equivalence classes of nodes of the uncompacted tree $Trie(text)$, where the equivalence means subtree isomorphism. In this sense, $val(v)$ is the longest representative of its equivalence class.

Let v be a node of $DAWG(text)$ distinct from the *root*. We define $suf[v]$ as the node w such that $val(w)$ is the longest suffix of $val(v)$ not equivalent to it. In other words, $val(w)$ is the longest suffix of $val(v)$ corresponding to a node different from v. Note that the definition implies that $val(w)$ is a proper suffix of $val(v)$. By convention, we define $suf[root] = root$. We also call the table suf the *table of suffix links*, and edges $(v, suf[v])$ *suffix links*. The suffix links on $DAWG(dababd)$ are presented in Figure 6.2. Since the value of $suf[v]$ is a word strictly shorter than $val(v)$, suf induces a tree structure on the set of nodes. The node $suf[v]$ is interpreted as the father of v in the tree. The tree of suffix links is the same as the tree structure of sets in Φ induced by the inclusion relation.

Theorem 6.1 *The number of nodes of $DAWG(text)$ is smaller than $2n$.*

Proof. Any two subsets of Φ are either disjoint or one is contained in the other. Thus, Φ has a tree structure (see Figure 6.2). Since the value of $suf[v]$ is a word strictly shorter than $val(v)$, the function suf induces a tree structure on the set of nodes. All leaves are pairwise disjoint subsets of $\{1, 2, \ldots, n\}$ (we do not count position 0 that is associated with the root, because it is not contained in any other *end-pos* set). Hence, there are at most n leaves. This does not directly imply the thesis because it can occur that some internal nodes have only one son (as in the example of Figure 6.2).

We partition nodes into two (disjoint) subsets according to the fact that $val(v)$ is a prefix of *text* or not. The number of nodes in the first subset is exactly $n + 1$ (number of prefixes of *text*). We now count the number of nodes in the other subset of the partition. Let v be a node such that $val(v)$ is not a prefix of *text*. Then $val(v)$ is a non-empty word that occurs in at least two different right contexts in *text*. But we can then deduce that at least two different nodes p and q (corresponding to two different factors of *text*) are such that $suf[p] = suf[q] = v$.

This shows that nodes like v have at least two sons in the tree inferred by suf. Since the tree has at most n leaves (corresponding to non-empty prefixes), the number of such nodes is less than n. Additionally note that if *text* contains two different letters, the root has at least two sons but cannot be counted in

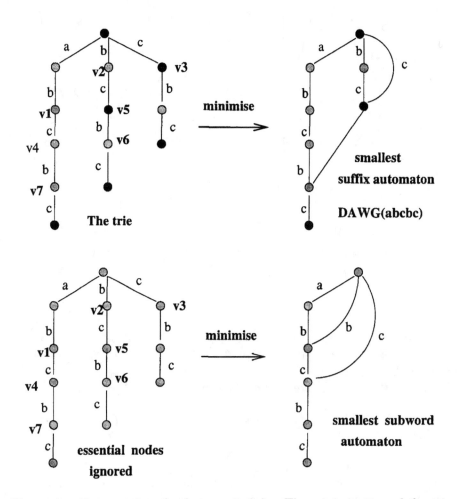

Figure 6.1: Representing the factors of *abcbc*. The minimization of the trie depends on whether we consider essential nodes or not. For example in the first case $v1$ and $v2$ are roots of non-isomorphic subtrees, while in the second case they are roots of isomorphic trees. Nodes $v7$ and $v6$ are roots of isomorphic subtrees in both cases. The roots of isomorphic subtrees are glued together. Usually DAWG's and smallest subword automata do not differ too much.

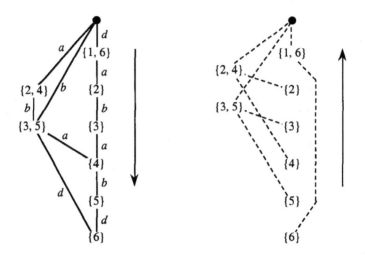

Figure 6.2: *DAWG(dababd)* and its suffix links. The tree of suffix links shows the structure of the family Φ of *end-pos* sets. The structure of suffix links of *dababd* is the same as the structure of the suffix tree of the reversed text *dbabad*.

the second subset because $val(root) = \varepsilon$ is prefix of *text*. If *text* is of the form a^n, the second subset is empty. Therefore, the cardinality of the class is indeed less than $n - 1$.

This finally shows that there are less than $(n + 1) + (n - 1) = 2n$ nodes.   □

Theorem 6.2 *DAWG(text) has less than $N + n - 1$ edges, where N is the number of its nodes. This is independent of the size of the alphabet.*

Proof. We consider a spanning tree T over *DAWG(text)*, and count separately the edges belonging to the tree and the edges outside the tree. The tree T is chosen to contain the branch labeled by the whole text. Since there are N nodes in the tree, there are $N - 1$ edges in the tree. Let us count the other edges of *DAWG(text)*. Let (v, w) be such an edge. We associate with it the suffix xay of *text* defined by: x is the label of the path in T going from the root to v, a is the label of the edge (v, w), y is any factor of *text* extending xa into a suffix of *text* for $x, y \in A^*$, $a \in A$.

The correspondence is one-to-one. The empty suffix is not considered, nor is *text* itself because it is in the tree.

It remains $n - 1$ suffixes, which is the maximum number of edges outside T. Hence the number of edges in *DAWG(text)* is less than $N + n - 1$.   □

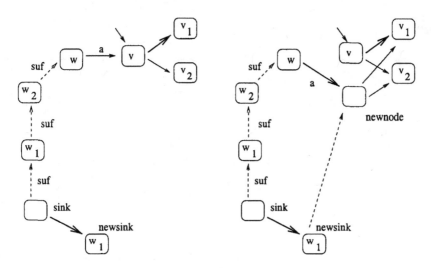

Figure 6.3: One iteration in the on-line construction of DAWG's. *newnode* is a clone of *v*.

The DAWG is not always strictly minimal. If minimality is understood in terms of finite automata (number of nodes, for short), then $DAWG(text)$ is the minimal automaton for the set of suffixes of *text*. Accepting states correspond to the essential states of the DAWG (nodes corresponding to suffixes).

6.2 On-line construction of subword graphs

An on-line linear-time algorithm for suffix DAWG computation can be viewed as a simulation of the algorithm *on-line-trie* from Chapter 4, in which isomorphic subtrees are identified. When the text is extended by the letter *a* to the right then the algorithm *on-line-trie* follows the *working path*: the sequence of suffix links until an edge labeled *a* is found if ever. Then the algorithm creates new vertices and updates suffix links. We describe a similar algorithm, but all leaves of the trie are now identified as a special node called *sink*. In addition the set of other nodes is partitioned into equivalent classes.

The peculiar feature of the on-line algorithm is that instead of gluing nodes together it either creates single new nodes from scratch or creates clones of existing nodes, so it makes a kind of state splitting that is opposite to gluing. This can be understood by the fact that there is a lot of gluing because all leaves are glued together, but later some of these leaves do not correspond to suffixes and should be split. State splitting is the basic operation of the algorithm.

Algorithm *on-line-DAWG*;
    create the one-node graph $G = DAWG(\varepsilon)$;
    $root := sink$; $suf[root] :=$ nil;
    for $i := 1$ to n **do begin**
        $a := text[i]$; create a new node *newsink*;
        make a solid a-edge $(sink, newsink)$; $w := suf[sink]$;
        while $(w \neq$ nil$)$ **and** $(son(w, a) =$ nil$)$ **do begin**
            make a non-solid a-edge $(w, newsink)$; $w := suf[w]$;
        end;
        $v := son(w, a)$;
        if $w =$ nil **then** $suf[newsink] := root$
        else if (w, v) is a solid edge **then** $suf[newsink] := v$
        else begin { split the node v }
            create a node *newnode* with the same outgoing edges as v,
            except that they are all non-solid;
            change (w, v) into a solid edge $(w, newnode)$;
            $suf[newsink] := newnode$; $suf[newnode] := suf[v]$;
            $suf[v] := newnode$; $w := suf[w]$;
            while $w \neq$ nil **and** (w, v) is a non-solid $a - edge$ **do begin**
                $\{*\}$ redirect this edge to *newnode*; $w := suf[w]$;
            end;
        end;
        $sink := newsink$;
    end

The algorithm processes the text from left to right. At each (Figure 6.4) step, it reads the next letter of the text and updates the DAWG built so far. In the course of the algorithm, two types of edges in the DAWG are considered: solid edges (thick in Figure 6.6) and non-solid edges. The solid edges are those contained in longest paths from the root. In other words, non-solid edges are those creating shortcuts in the DAWG. The adjective *solid* means that once these edges are created, they are not modified during the rest of the construction. On the contrary, the target of non-solid edges may change after a while. Figures 6.5, 6.8 and 6.7 show a run of the algorithm on *text* = *aabbab*. The schema of one stage of the algorithm is graphically presented in Figure 6.3.

The transformation of *DAWG(aabba)* into *DAWG(aabbab)* points out a crucial operation of the algorithm. The addition of letter b to *aabba* introduces new factors to the set $\mathcal{F}(aabba)$. They are all suffixes of *aabbab*. In

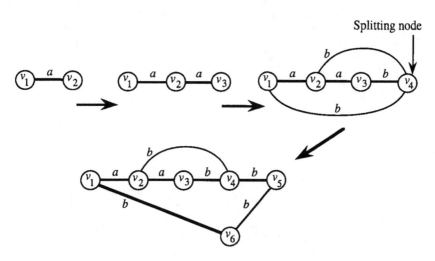

Figure 6.4: Iterative computation of $DAWG(aabb)$. Solid edges are thick.

$DAWG(aabba)$, nodes corresponding to suffixes of $aabba$, namely, v_1, v_2, v_7 in Figure 6.7 and 6.8, will now have an out-going b-edge. Consider node v_2 in $DAWG(aabba)$. It has an outgoing non-solid b-edge. The edge is a shortcut between v_2 and v_4, compared to the longest path from v_2 to v_4 which is labeled by ab. The two factors ab and aab are associated with the node v_4. But in $aabbab$, only ab becomes a suffix, not aab. This is the reason why the node v_4 is split into v_4 and v_9 in $DAWG(aabbab)$.

When splitting a node, it may also occur that some edges need to be redirected to the created node. This situation is illustrated in Figure 6.5 by $DAWG(aabbabb)$ (see also Figure 6.6). In $DAWG(aabbab)$ (Figure 6.5), the node v_5 corresponds to factors $bb, abb,$ and $aabb$. In $aabbabb$, only bb and abb are suffixes. Hence, in $DAWG(aabbabb)$, paths labeled by bb and abb should reach the node v_{11}, a clone of node v_5 obtained by the splitting operation. In the algorithm, we denote the son v of node w by $son(w, a)$ such that $label(w, v) = a$. We assume that $son(w, a) = $ nil if there is no such node v. For each node of $DAWG(text)$ a suffix link named suf is defined. It creates lists of *working paths* as shown in Figure 6.3. There, the working path is $w_1 = suf[sink]$, $w_2 = suf[w_1]$, $w = suf[w_2]$. Generally, the node w is the first node on the path having an outgoing edge (w, v) labeled by letter a. If this edge is non-solid, then v is split into two nodes: v itself, and *newnode*. The latter is a clone of node v, in the sense that out-going edges and the suffix-pointer for *newnode* are the same as for v. The suffix link of *newsink* is set to *newnode*. Otherwise, if edge (w, v) is solid, the only action performed at this step is to set the suffix

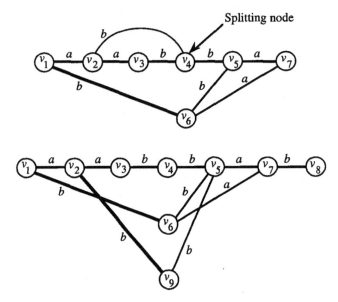

Figure 6.5: Transformation of $DAWG(aabba)$ into $DAWG(aabbab)$. The node v_9 is a clone of v_4.

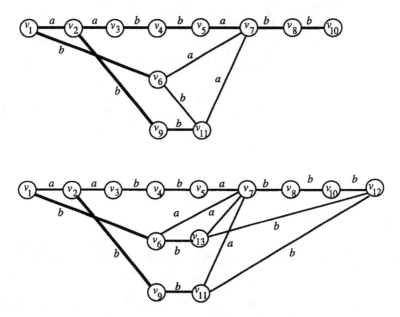

Figure 6.6: The suffix DAWG's of *aabbabb* and *aabbabbb*. The working path in the first graph consists of nodes v_{11} and v_6, since $suf[sink] = v_{11}$ and $suf[v_{11}] = v_6$. There is no edge labeled by b from v_{11} so a b-edge is added to v_{11}. The non-solid edge labeled by b points from v_6 to v_{11}, a clone v_{13} of v_{11} is created and this edge is redirected to the clone.

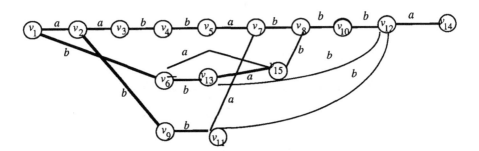

Figure 6.7: The suffix DAWG of *aabbabbba*. The node v_{15} is a clone of v_7. Several edges leading to the cloned node are redirected. The non-solid a-edge $v_6 \to v_7$ is redirected to v_{15}.

link of *newsink* to v.

Theorem 6.3 *Algorithm on-line-DAWG computes DAWG(text) in linear time.*

Proof. We estimate the complexity of the algorithm. We proceed similarly as in the analysis of the suffix tree construction. Let $sink_i$ be the sink of $DAWG(text[1..i])$. What we call the working path consists of nodes $w_1 = suf[sink]$, $w_2 = suf[w_1]$, ..., $w_{k+1} = suf[w_k]$. The complexity of one iteration is proportional to the length k, of the working path plus the number k' of redirections made at step * of the algorithm. Let K_i be the value of $k + k'$ at the i-th iteration. Let $depth(u)$ be the depth of node u in the graph of suffix links corresponding to the final DAWG (the depth is the number of applications of *suf* needed to reach the root). Then, the following inequality can be proved (we leave it as an exercise): $depth(sink_{i+1}) \leq depth(sink_i) - K_i + 2$, and this implies

$$K_i \leq depth(sink_i) - depth(sink_{i+1}) + 2.$$

Hence, the sum of all K_i's is linear. $\square$

6.3 The reverse perspective

There are close relations between suffix trees and subword graphs, since they are compact representations of the same trie. We examine the relations between $DAWG(w)$ and $ST(w^R)$ and principally the facts that suffix links of $DAWG(w)$ correspond to $ST(w^R)$, and conversely suffix links of $ST(w^R)$ give solid edges of $DAWG(w)$. Two factors of *text*, x and y, are equivalent iff their *end-pos* sets are equal. This means that one of the words is a suffix of the other (say y is a suffix of x), and then wherever x appears then y also appears in text. However, what happens if, instead of end-positions, we consider start-positions? Let us look from the "reverse perspective" and look at the reverse pattern. Suffixes then become prefixes, and end-positions become first-positions. Denote by *first-pos*(x) the set of first positions of occurrences of x in *text*. Recall that a chain in a tree is a maximal path in which all the nodes have out-degree one and they are non-essential nodes except the last one. The following obvious lemma is crucial for understanding the relationship between suffix trees and DAWG's.

Lemma 6.1 *The following three properties are equivalent for two factors x and y of text:*
(1) end-pos(x) = end-pos(y) in text,
(2) first-pos(x^R) = first-pos(y^R) in text^R,
(3) x^R, y^R are contained in the same chain of the uncompacted tree Trie$(text^R)$.

Observation. The reversed values of nodes of $ST(text^R)$ are the longest representatives of equivalence classes of factors of *text*. Hence nodes of $ST(text^R)$ can be treated as nodes of $DAWG(text)$.

From Suffix trees to subword graphs.

We strongly require here that all suffixes of the text have corresponding nodes (essential nodes) in the suffix tree T, though some of these nodes can be of outdegree one. Actually these are the trees constructed by Ukkonen algorithm, which gives also suffix links. We use another family of links which are related to suffix links (but work in reversed direction), the extension links denoted by $ext[a, v]$. The value of $ext[a, v]$ is the node w for which $val(w)$ is the shortest word having prefix ax, where $x = val(v)$. If there is no such node w, then $ext[a, v] = $ nil.

Theorem 6.4 *(1) $DAWG(text)$ is the graph of extension links of the suffix tree $T = ST(text^R)$.*
(2) Solid edges of $DAWG(text)$ are reverses of suffix links of T.

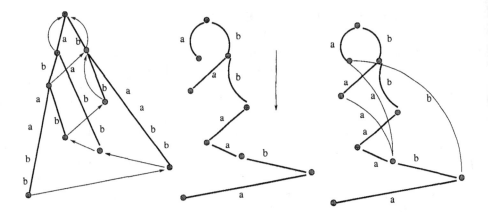

Figure 6.8: **(1)** The suffix tree of *abaabb* with suffix links (cutting first letter).
(2) Reverses of suffix links give solid edges of the DAWG of the reversed text
bbaaba. **(3)** Three additional non-solid edges corresponding to extension links
of the suffix tree are added to obtain the complete DAWG of *bbaaba*.

Proof. The nodes of the DAWG correspond to nodes of T. In the DAWG
for *text*, the edge labeled a goes from the node v with $val(v) = x$ to node w
with $val(w) = y$ iff y is the longest representative of the class of factors that
contain word xa. If we consider the reversed text, then it means that ax^R is
the longest word y^R such that $first\text{-}pos(y^R) = first\text{-}pos(ax^R)$ in $text^R$. This
exactly means that

$$y^R = ext[a, x^R] \text{ and } link[a, x^R] = y^R.$$

In conclusion, this also means that table *link* gives exactly all solid edges of
$DAWG(text)$.   □

Theorem 6.5 *If we are given a suffix tree T with the table of suffix links then
the table of extension links can be computed in linear time.*

Proof. We reverse extension links: if $suf[a \cdot \alpha] = \alpha$ then $ext[\alpha, a] = a \cdot \alpha$,
where we identify nodes with strings. Afterwards, we process the tree bottom
up:

    if $ext[u, a] = $ nil **and** $ext[w, a] \neq $ nil for a son w of u **then**
        $ext[u, a] := ext[w, a]$.

The whole process obviously takes linear time for constant-size alphabets.   □

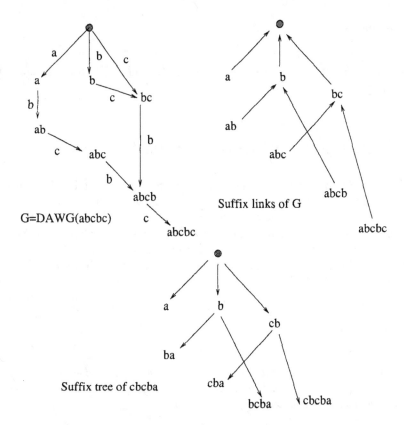

Figure 6.9: The tree of suffix-links of $DAWG(abcbc)$ is the suffix tree $\mathcal{ST}((abcbc)^R)$. The nodes are identified with their string representatives (in the DAWG these are the labels of the longest paths leading to nodes).

Observation. If we have a suffix tree with essential nodes known, but without suffix links then suffix links, can be easily computed bottom-up in linear time.

From subword graphs to suffix trees.

Let us look at the opposite direction. Assume again that we identify nodes with their corresponding string representations. For each suffix link $suf[y] = x$ in $DAWG(w)$ create the edge $x^R \to y^R$ labelled $y^R \, // \, x^R$, where $//$ means the operation of cutting the prefix. The construction is illustrated in Figure 6.9 on an example DAWG (only suffix links of the DAWG are shown). The same arguments as used in the proof of Theorem 6.4 can be applied to show the following fact.

Theorem 6.6 *The tree of reversed suffix links of DAWG(text) is the suffix tree $T(text^R)$.*

6.4 Compact subword graphs

DAWG's result by identifying isomorphic subtrees in suffix tries. We consider the construction of DAWG's by identifying isomorphic subtrees in suffix trees. This construction gives a succinct representation of $DAWG(w)$ called the *compact DAWG* od w and demoted as $CDAWG(w)$. This is a version of a DAWG in which edges are labeled by words, and there are no non-essential nodes of outdegree one (chains of non-essential nodes are compacted). In other words this is a *compacted DAWG*. It is very easy to convert $CDAWG(w)$ into $DAWG(w)$ and vice versa. We show the transformation of a suffix tree into a CDAWG. The basic procedure is the computation of equivalent classes of subtrees is based on the following classical algorithm (see [AHU 74] for example).

Lemma 6.2 *Let T be a rooted ordered tree in which the edges are labeled by letters (assumed to be of constant size). Then, isomorphic classes of all subtrees of T can be computed in linear time.*

We illustrate the algorithm with the example text string $w = baaabbabb$. For technical reasons the end-marker $\#$ is added to the word w. But this end-marker is later ignored in the constructed DAWG (it is only needed at this stage). Then, the equivalence classes of isomorphic subtrees are computed. The roots of isomorphic subtrees are identified, and we get the compacted version G' of $DAWG(w)$. The difference between the actual structure and the DAWG is related to the lengths of strings that are labels of edges. In the DAWG each edge is labeled by a single symbol. The "naive" approach could be to break each edge labeled by a string of length k down into k edges. But the resulting graph could have a quadratic number of nodes. We apply such an approach with the following modification. By the weight of an edge we mean the length of its label. For each node v we compute the heaviest (with the largest weight) incoming edge. Denote this edge by $inedge(v)$. Then, for each node v, we perform a local transformation $local-action(v)$ (see Figure 6.12). It is crucial that all these local transformations $local-action(v)$ are independent and can be performed for all v. The entire process takes linear time.

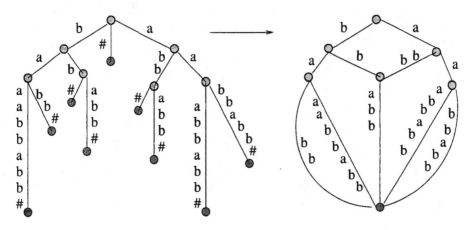

Figure 6.10: After identification of isomorphic classes of nodes of the suffix tree and removal of end-marker # we obtain the compact DAWG for *baaabbabb*.

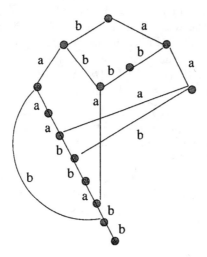

Figure 6.11: After decompacting the compacted DAWG of Figure 6.10 we get the suffix DAWG of *baaabbabb*.

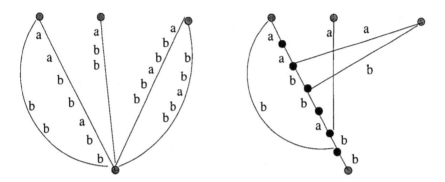

Figure 6.12: The decompaction consists of local transformations. New nodes (black nodes) are introduced on the heaviest incoming edges. It is illustrated for all edges arriving to the sink node, see Figure 6.11.

Bibliographic notes

The impressive feature of DAWG's is their linear size, first discovered in [BBEHC 83]. After marking nodes of the DAWG associated with suffixes of the text as terminal states, the DAWG becomes an automaton recognizing the suffixes of the text. In fact, this is indeed the minimal automaton for the set of suffixes of the text. This point is discussed in [Cr 85] where an algorithm to build the minimal automaton recognizing the set of all factors of the text is also presented (the factor automaton can be slightly smaller than $DAWG(text)$). Constructions of $DAWG(text)$ by Blumer et al. [BBEHCS 85] and Crochemore [Cr 86] are essentially the same.

On-line construction of DAWG's can be treated as a simulation of Ukkonen algorithm, in which isomorphic classes of vertices are nodes of the DAWG. The standard application of DAWG's is for building efficient dictionaries of factors of many strings, see [BBHME 87].

The relationship between DAWG's and suffix trees is adeptly described in [CS 85]. The application of DAWG's to find the longest common factor of two words is presented in [Cr 86] (see also [Cr 87]).

Chapter 7

Text algorithms related to sorting

The numbering or naming of factors of a text, corresponding to the sorted order of these factors, can be used to build a useful data structure. In particular, the sorted sequence of all suffixes has a similar asymptotic efficiency for many problems as that of suffix trees. Most algorithms in this chapter are optimal within a logarithmic factor, but they are easy to implement and easier to understand, compared with asymptotically more efficient and at the same time much more sophisticated algorithms for DAWG's and suffix trees. Linear-time algorithms for sorting integers in the range $[1..n]$ can be successfully used in several text algorithms, some of them are presented below.

7.1 The naming technique: *KMR* algorithm

The central notion used in the algorithms of the section is called *naming* or *numbering*. We define a version of the Karp-Miller-Rosenberg algorithm (*KMR* algorithm) as an algorithm computing the data structure called the *dictionary of basic factors*. In the algorithm we assign names to certain subwords, or pairs of subwords.

Assume we have a sequence
$$S = (s_1, s_2, \ldots, s_t)$$
of at most n different objects. The naming of S is a table
$$X[1], X[2], \ldots, X[t]$$
that satisfies conditions (1-2) below. If, in addition, it satisfies the third condition then the naming is called a *sorted naming*.

Positions	=	1	2	3	4	5	6	7	8			
text	=	a	b	a	a	b	b	a	a	#	#	#
$Name_1$	=	1	2	1	1	2	2	1	1			
$Name_2$	=	2	4	1	2	5	4	1	3			
$Name_4$	=	3	6	1	4	8	7	2	5			
Name of factor	=	1	2	3	4	5	6	7	8			
Pos_1	=	1	2									
Pos_2	=	3	1	8	2	5						
Pos_4	=	3	7	1	4	8	2	6	5			

Figure 7.1: The dictionary of basic factors for an example text: tables of k-names and of their positions. The k-name at position i corresponds to the factor $text[i..i + k - 1]$; its name is its rank according to lexicographic order of all factors of length k (order of symbols is $a < b < \#$). Indices k are powers of two. The tables can be stored in $O(n \log n)$ space.

1. $s_i = s_j \Leftrightarrow X[i] = X[j]$, for $1 \le i, j \le t$.

2. $X[i] \in [1..n]$ for each position i, $1 \le i \le t$.

3. $X[i]$ is the rank of s_i in the ordered list of the different elements of S.

Given the string $text$ of length n, we say that two positions are k-equivalent if the factors of length k starting at these positions are equal. Such an equivalence is best represented by assigning to each position i a name or a number to the factor of length k starting at this position. The name is denoted by $Name_k[i]$ and called a k-name. We assume that the table $Name_k$ is a good and sorted numbering of all factors of a given length k.

We consider only those factors of the text whose length k is a power of two. Such factors are called **basic factors**. The *name of a factor* is denoted by its rank in the lexicographic ordering of factors of a given length. For each k-name r we also require (for further applications) a link $Pos_k[r]$ to any one position at which an occurrence of the k-name r starts. Symbol $\#$ is a special end-marker that has the highest rank in the alphabet. The text is padded with enough end-markers to let the k-name at position n defined.

The tables $Name$ and Pos for a given text w are together called its **dictionary of basic factors** and is denoted by $DBF(w)$. This dictionary is the basic data structure of the chapter.

Remark. String matching for $text$ and pattern pat of length m can be easily reduced to the computation of a table $Name_m$. Consider the string $w = pat \& text$, where $\&$ is a special symbol not in the alphabet. Let $Name_m$ be the

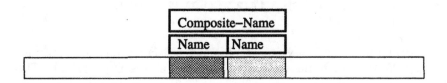

Figure 7.2: The object is decomposed into a constant number of parts of identical shape; its composite name consists in assembling names of its parts. The function *Sort-Rename* converts composite names into integers.

array which is part of $DBF(pat\&text)$. If $q = Name_m[1]$ then the pattern *pat* starts at all positions i on *text* such that $Name_m[i + m + 1] = q$.

Figure 7.1 displays $DBF(abaabbaa\#)$. Three additional $\#$'s are appended to guarantee that all factors of length 4 starting at positions $1, 2, \ldots, 8$ are well defined. The figure presents tables *Name* and *Pos*. In particular, the entries of Pos_4 give the lexicographically-sorted sequence of factors of length 4. This is the sequence of factors of length 4 starting at positions $3, 7, 1, 4, 8, 2, 6, 5$. The ordered sequence is:

$$aabb, \quad aa\#\#, \quad abaa, \quad abba, \quad a\#\#\#, \quad baab, \quad baa\#, \quad bbaa.$$

The central machinery of *KMR* algorithm is the procedure *Sort-Rename* that is defined now. Let S be a sequence of total size $t \leq n$ containing elements of some linearly ordered set. The output of *Sort-Rename(S)* is an array of size t, which is a good and sorted naming of S.

Example. Let $S = (ab, aa, ab, ba, ab, ba, aa)$. Then

$$Sort\text{-}Rename(S) = (2, 1, 2, 3, 2, 3, 1)$$

For a given k, define (see Figure 7.2),

$$Composite\text{-}Name_k[i] = (Name_k[i], Name_k[i + k]).$$

KMR algorithm is based on the following simple property of naming tables.

Lemma 7.1 [Key-Lemma] $Name_{2k} = Sort\text{-}Rename(Composite\text{-}Name_k)$.

The main part of algorithm *Sort-Rename* is the lexicographic sort. We explain the action of *Sort-Rename* on the following example:

$$x = ((1,2),(3,1),(2,2),(1,1),(2,3),(1,2)).$$

The method to compute a vector X of names satisfying conditions (1-3) is as follows. We first create the vector y of composite entries $y[i] = (x[i], i)$. Then, the entries of y are lexicographically sorted. Therefore, we get the ordered sequence

$$((1,1),4),((1,2),1),((1,2),6),((2,2),3),((2,3),5),((3,1),2).$$

Next, we partition this sequence into groups of elements having the same first component. These groups are consecutively numbered starting from 1. The last component i of each element is used to define the output element $X[i]$ as the number associated with the group. Therefore, in the example, we ge

$$X[4] = 1, \ X[1] = 2, \ X[6] = 2, \ X[3] = 3, \ X[5] = 4, \ X[2] = 5.$$

The linear-time lexicographic sorting based on bucket sort can be applied (see [AHU 83], for example). Doing so, the procedure *Sort-Rename* has the same complexity as sorting n elements of a special form, which yields the following statement.

Lemma 7.2 *If the vector x has size n, and its components are pairs of integers in the range $(1, 2, \ldots, n)$, Sort-Rename(x) can be computed in time $O(n)$.*

The dictionary of basic factors is computed by the algorithm *KMR* above. Its correctness results from fact (∗) below. The number of iterations is logarithmic, and the dominating operation is the call to procedure *Sort-Rename*.

Once all vectors $Name_p$, for all powers of two smaller than r, are computed, we easily compute the vector $Name_q$ in linear time for each integer $q < r$. Let t be the greatest power of two not greater than q. We can compute $Name_q$ by using the following fact:

(∗) $Name_q[i] = Name_q[j]$ iff $(Name_t[i] = Name_t[j])$ and $(Name_t[i + q - t] = Name_t[j + q - t])$.

Algorithm *KMR*;
{ a version of the Karp-Miller-Rosenberg algorithm }
{ Computation of *DBF(text)* }
$K :=$ largest power of 2 not exceeding n;

Computation of table *Name*:
$Name_1 := Sort\text{-}Rename(text)$;
for $k := 2, 4, \ldots, K$ **do**
$Name_{2k} := Sort\text{-}Rename(Composite\text{-}Name_k)$;

Computation of table *Pos*:
for $k = 1, 2, 4, \ldots, K$ **do**
for $1 \le i \le n$ **do** $Pos_k[Name_k[i]] := i$;

Let us denote by *LongestRepFactor(text)* the length of the *longest repeated factor* of *text*. It is the longest word occurring at least twice in *text* (occurrences are not necessarily consecutive). When there are several such longest repeated factors, we consider any of them. Let also *LongestRepFactor$_k$(text)* be the maximal length of the factor that occurs at least k times in *text*. Let us denote by $REP_k(r, text)$ the function that tests is there is a k-repeating factor of size r. Such function can be easily implemented to run in linear time if we have *DBF(w)*.

Theorem 7.1 *The function LongestRepFactor$_k$(text) can be computed in time $O(n \log n)$ for alphabets of size $O(n)$.*

Proof. We can assume that the length n of the text is a power of two; otherwise a suitable number of "dummy" symbols are appended to the text. The algorithm *KMR* is used to compute *DBF(text)*. We then apply a kind of binary search using function $REP_k(r, text)$: the binary search looks for the maximum r such that $REP_k(r, text) \ne$ nil. If the search is successful then we return the longest (k times) repeated factor. Otherwise, we report that there is no such repetitions. The sequence of values of $REP_k(r, text)$ (for $r = 1, 2, ., n - 1$) is "monotonic" in the following sense: if $r1 < r2$ and $REP_k(r2, text) \ne$ nil, then $REP_k(r1, text) \ne$ nil. The binary search behaves similarly to searching an element in a monotonic sequence. It has $\log n$ stages; at each stage the value $REP_k(r, text)$ is computed in linear time. Altogether the computation takes $O(n \log n)$ time. This completes the proof. $\square$

The longest repeated factor problem for texts can also be solved in a straightforward way if we have already constructed the suffix tree $ST(tree)$

(see Chapters 4 and 5). The value $LongestRepFactor(text)$ is the length of the longest path (in the sense of the length of word corresponding to the path) leading from the root of $ST(tree)$ to an internal node. Generally, $LongestRepFactor_k(text)$ is the length of a longest path leading from the root to an internal node in which the subtree contains at least k leaves. The computation of such a path can be accomplished easily in linear time. The preprocessing needed to construct the suffix tree makes the whole algorithm much more complicated than the one applying the strategy of KMR algorithm. Nevertheless, this proves that the computation takes linear time with suffix trees.

7.2   Two-dimensional KMR algorithm

In the case of arrays, basic factors are $k \times k$ sub-arrays, where k is a power of two. In this situation, $Name_k[i,j]$ is the name of the $k \times k$ sub-array of a given array $text$ having its upper left corner at position (i,j). We have primarily discussed the construction of dictionaries of basic factors for strings. The construction in the two-dimensional case is a simple extension of that used for one-dimensional data. In the two-dimensional case, there is a fact analogous to $(*)$, which makes the algorithm work similarly.

The naming technique is used on sub-arrays. Let $Name_r[i,j]$ be the number associated with the $r \times r$ sub-array of the array T having its upper-left corner at position (i,j). There is a fact, analogous to $(*)$, illustrated by Figure 7.3:

$(**)$ $Name_{2p}[i,j] = Name_{2p}[k,l]$ iff the following conditions are satisfied:
$Name_p[i,j] = Name_p[k,l];$ $Name_p[i+p,j] = Name_p[k+p,l];$
$Name_p[i,j+p] = Name_p[k,l+p];$
$Name_p[i+p,j+p] = Name_p[k+p,l+p].$

The longest repeated factor problem generalizes an equivalent 2-dimensional problem in a straightforward way. For this problem, KMR algorithm gives $O(N \log N)$ time complexity, which is the best upper bound known up to now. The algorithm works also for finding repetitions in trees as well.

Using fact $(**)$, the whole computation of repeating $2p \times 2p$ sub-arrays reduces to the computation of repeating $p \times p$ sub-arrays. The matrix $Name_{2p}$ is computed from $Name_p$ using a procedure analogous to $Sort\text{-}Rename$. Here, the internal lexicographic sorting is executed on elements having four components (instead of two for texts). We then get the next result.

Theorem 7.2 *The size of a longest repeated sub-array of an $n \times n$ array of symbols can be computed in $O(N \log N)$ time, where $N = n^2$.*

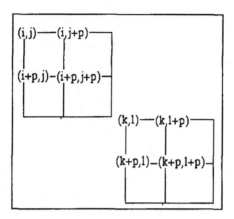

Figure 7.3: A repeated sub-array of size $2p \times 2p$. The occurrences can overlap. The name of a sub-array is determined by names of its four quadrants.

7.3 Suffix arrays

There is a clever and rather simple way to deal with all (non-empty) suffixes of a text: to arrange their list in increasing lexicographic order with the aim of performing binary searches on them. The implementation of this idea leads to a data structure called a *suffix array*. It is not exactly a "good" representation, in the sense defined at the beginning of Chapter 4. But it is "almost good." This means that it satisfies the following conditions:

(1) it has $O(n)$ size,

(2) it can be constructed in $O(n \log n)$ time,

(3) the query $x \in \mathcal{F}(text)$ can be answered in $O(|x| + \log n)$ time.

So the time required to construct and use the structure is slightly greater than that needed to compute the suffix tree (it is $O(n \log |A|)$ for the latter). But suffix arrays have two advantages:

- their construction is rather simple; it is even commonly admitted that, in practice, it behaves better than the construction of suffix trees,

- it consists of two linear-size arrays which, in practice again, take little memory space (typically three times less space than suffix trees).

Let $text = a_1 a_2 \ldots a_n$, and let $p_i = a_i a_{i+1} \ldots a_n$ be the i-th (non-empty) suffix of the text. Let $SufPos[k]$ be the position i where the k-th smallest suffix

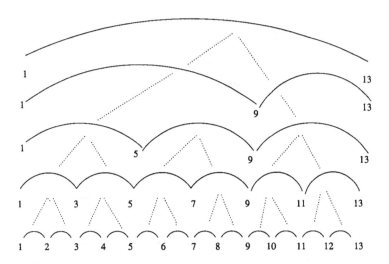

Figure 7.4: The edges correspond to regular pairs, these are pairs which take part in a binary search.

of *text* starts (according to the lexicographic order of the suffixes). In other words, the k-th smallest suffix of *text* is $p_{SufPos(k)}$. Denote by $LCPref(u, w)$ the length of the longest common prefix of words (u, w).

We say that a pair of positions (i, j) in $[1 .. n]$ is *regular* iff there exist integers p, k such that $i = 1 + p \cdot 2^k$ and $j = \min\{n, i + (p + 1) \cdot 2^k\}$. The set of regular pairs is illustrated in Figure 7.4 for $n = 13$. In particular $(1, n)$ is always a regular pair. The structure of regular pairs is a binary search tree which enables to find a pattern using recursion of logarithmic depth.

Let $Suffix(k)$ be the lexicographically k-th suffix of text; it starts at position $SufPos[k]$. The *suffix array* of *text* is the data structure consisting of both the array $SufPos$, and the values

$$LCP[i, j] = LCPref[Suffix(i), Suffix(j)],$$

for all regular pairs (i, j). The entire data structure has $O(n)$ size because the number of regular pairs is linear, so it satisfies condition (1). To show that condition (2) is satisfied we can use the dictionary of basic factors to build the arrays. Note that condition (2) is rather intuitive because it involves sorting n words having certain mutual dependencies. But, in our opinion, the most interesting property of *suffix arrays* is that they satisfy condition (3).

Theorem 7.3 *Assume that the suffix array of text is computed (we have tables SufPos and LCPref for regular pairs). Then, for any word x, we can answer the query "$x \in \mathcal{F}(text)$" in $O(|x| + \log |text|)$ time.*

Remark. The lexicographic ordering of suffixes has the following nice property. Let

$$min_x = \min\{k : x \text{ is a prefix of } p_{SufPos[k]}\},$$
$$max_x = \max\{k : x \text{ is a prefix of } p_{SufPos[k]}\}.$$

Then all occurrences of the subword x in *text* are at positions listed in the table *SufPos* between indices min_x and max_x.

We present an algorithm, written as a recursive binary search function *find*, which computes an occurrence of the pattern. It is a binary search for x inside the sorted sequence of suffixes. Let us define

$$Suffix[i,j] \;=\; (Suffix(i), Suffix(i+1), \ldots , Suffix(j)).$$

We describe the pattern searching recursively. The value of *Search(Left, Right)* is the position of a suffix in the interval of suffixes *Suffix*[*Left, Right*] having x as a prefix. If there is no such suffix then the output is nil. For a regular pair (i, j) of non-consecutive positions, denote by $Middle(i, j)$, *middle position* between i and j, the position such that the pairs $(i, Middle(i, j))$ and $(Middle(i, j), j)$ are regular pairs. If i, j are consecutive then $Middle(i, j) = i$.

The crucial component of the algorithm is an additional *memory* given by an integer M. The **basic invariant** is:

$M = LCPref(Suffix(Left), x)$ or $M = LCPref(Suffix(Right), x)$;
the value of M does not decrease during a run of the algorithm.

The basic additional function is $Compare(x, Mid)$. The function compares, as strings, x with $Suffix(Mid)$, and returns the value *left* or *right* depending on whether x is lexicographically smaller or greater than $Suffix(Mid)$. A naive comparison would consist in comparing consecutive letters starting from the first one. However the clever comparison uses the knowledge of M and saves usually a lot of single letter comparisons. If a letter at some position in x is compared and there is an equality of two corresponding letters then this position in x will never be checked again, since the value of M will increase. Only letters at positions larger than M are possibly checked later.

Let us denote $Suffix(Mid) = u$, $Suffix(Right) = w$. Figure 7.5 illustrates the situation where $M = LCPref(x, w)$. The situation where $M = LCPref(x, Suffix(Left)))$ is symmetric.

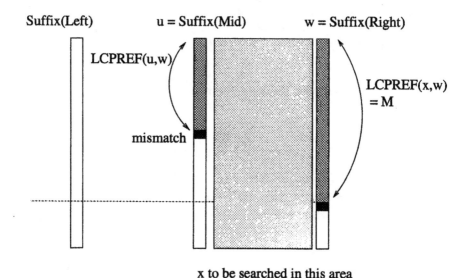

x to be searched in this area

Figure 7.5: Suffix array searching: the case when $LCPref(u, w) < M$.

Description of the function *Compare*. There are two main possible cases:

Case1: $LCPref(u, w) < M$ (the most interesting case).
We know that x is not a prefix of $Suffix(Mid)$ and x should be searched in the interval $[Mid, \dots, Right]$, see Figure 7.5. The function *Compare* returns the value *right*.

Case2: $LCPref(u, w) \geq M$.
We compare letters of x against letters of u starting from position $M + 1$ until the first mismatch occurs, or until the whole x or u is read. We update M, by adding the number of matching positions. If there is a mismatch, and the letter of x is smaller than the corresponding letter of u then we know that x should be searched in the interval $[Left, \dots, Mid]$; if there is a mismatch with the opposite result, or if u is a proper prefix of x, then it should be searched in $[Mid, \dots, Right]$. In the first case we return *left*, in the second case the output is *right*. In the third case, when there is no mismatch and $M = |x|$, we assume that *Compare* returns the value *left*.

Instead of searching the pattern in the original text we search it in the sorted list of suffixes. Initially, $M = LCPref(Suffix(1), x)$. We call $Search(1, n)$. If the output is i, then the pattern x starts at position i in the text.

Theorem 7.4 *The data structure consisting of two tables SufPos and LCP has the following properties: it has $O(n)$ size; it can be constructed in $O(n \log n)$ time; each query $x \in \mathcal{F}(text)$ can be answered in $O(|x| + \log n)$ time; the set of all r occurrences of x in text can be found in time $O(|x| + \log n + r)$.*

Proof. The first two points follow from the efficient use of the dictionary of basic factors. The third point follows from the analysis of the function *Search*. We make logarithmic number of iterations, and the total time spent in making symbols comparisons when calling *Compare* is $O(|x|)$, due to the savings implied by the use of variable M.

If we look carefully at the function *Search* then it happens that the returned value i is the index of the lexicographically first suffix that has x as a prefix. This is due to the fact that in our description of the function *Compare* we agreed that in the situation where there is no mismatch and $M = |x|$, *Compare* returns the value *left*. If we change here *left* to *right* then $Search(1, n)$ returns the index of the lexicographically last suffix containing x as a prefix. All occurrences of x are among these two suffixes and they can be listed from the table *SufPos*. This additionally proves the last point. □

---

function *Search*(*Left*, *Right*): integer;
{ intelligent binary search for x in the sorted sequence of suffixes }
    { (*Left*, *Right*) is a regular pair }
    if *Left* = *Right* **and** $M = |x|$ **then return** *SufPos*[*Left*]
    else if *Left* = *Right* **and** $M < |x|$ **then return** nil
    else begin
        Mid := *Middle*(*Left*, *Right*);
        if *Compare*(*x*, *Mid*) = left **then return** *Search*(*Left*, *Mid*)
        else if *Compare*(*x*, *Mid*) = right
        then return *Search*(*Mid*, *Right*)
    end

7.4 Constructing suffix trees by sorting

One of the most important algorithms in stringology is Farach's suffix-tree construction. It works in linear time, independently of the size of the alphabet, but assuming the letters are integers. Unfortunately Farach's algorithm is still

p_4 a a b a b a b b a #

p_2 a b a a b a b a b b a #

p_5 a b a b a b b a #

p_7 a b a b b a #

p_9 a b b a #

p_{12} a #

p_3 b a a b a b a b b a #

p_1 b a b a a b a b a b b a #

p_6 b a b a b b a #

p_8 b a b b a #

p_{11} b a #

p_{10} b b a #

Figure 7.6: The sorted sequence (top-down) of 12 suffixes of the string $x = babaabababba\#$ with the *lcp-values* between suffixes (length of shaded straps). The special suffix # and the empty suffix are not considered. We have: $lcp = [1, 3, 4, 2, 1, 0, 2, 4, 3, 2, 1]$ and $SufPos = [4, 2, 5, 7, 9, 12, 3, 1, 6, 8, 11, 10]$.

too complicated. To show a flavor of Farach's algorithm we show only an algorithm which has a similar complexity assuming that we already know the sorted order of all suffixes p_1, p_2, ... of the string x, where $p_i = x_i x_{i+1} \ldots x_n$ (though computing this order is the hardest part of Farach's algorithm). We need a portion of the *LCPref* table, corresponding only to pairs of adjacent suffixes (in the sorted sequence):

$$lcp[i] = LCPref(p_{SufPos[i]}, \; p_{SufPos[i+1]}).$$

In other words, $lcp[i]$ is the length of the common prefix of the i-th and $(i+1)$-th smallest suffixes in the lexicographic order. The table is illustrated, on an example string, in Figure 7.6. We show constructively the following theorem.

Theorem 7.5 *Assume we know the table SufPos[k] of a given text x. Then the suffix tree for x can be computed in linear time, independently of the size of the alphabet.*

Proof. Assume first we know the table *lcp* and the sorted sequence of suffixes is:

$$p_{i_1}, p_{i_2}, p_{i_3}, \ldots, p_{i_n}.$$

We create the suffix tree by inserting consecutive suffixes: each time we add one edge outgoing the rightmost branch of the tree. The main operation is to find a *splitting node* v on this branch. When the tree T_k is created after inserting $p_{i_1}, p_{i_2}, p_{i_3}, \ldots p_{i_k}$, then the splitting node can be found on the rightmost branch at total depth $lcp[k]$. To find this node we follow the rightmost branch bottom-up starting at the rightmost leaf, and *jumping* on each (compacted) edge.

The total work is amortized by the total decreases in the depth of the rightmost leaf. We distinguish here between the depth (number of nodes) and total depth (length of the string corresponding to a path from the root). □

The algorithm shortly described in the previous proof is called here the *suffixes-insertion* algorithm. The history of the algorithm is shown in Figures 7.7 and 7.8. This is probably the simplest suffix tree construction algorithm, but it assumes we have sorted the sequence of suffixes. This can be easily done in $O(n \log n)$ time with KMR algorithm, but needs a sophisticated algorithm to be done in $O(n)$ time independently of the alphabet size (alphabet is a set of integers).

There is an interesting algorithm, which computes the table *lcp* in a rather strange manner. Let $rank(i)$ be the rank of p_i in the lexicographic ordering. Assume $lcp[0] = 0$. Then we compute

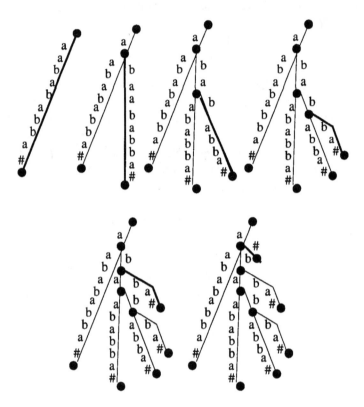

Figure 7.7: The first 6 stages performed by the algorithm *suffixes-insertion* to build the suffix tree of *babaabababba*# (see Figure 7.6). At each stage, one edge is added (indicated by a thick line) from a node v on the rightmost path, corresponding to the last processed i-th smallest suffix, to the leaf corresponding to $(i+1)$-th smallest suffix. The total depth of the *splitting node* v equals $lcp[i]$.

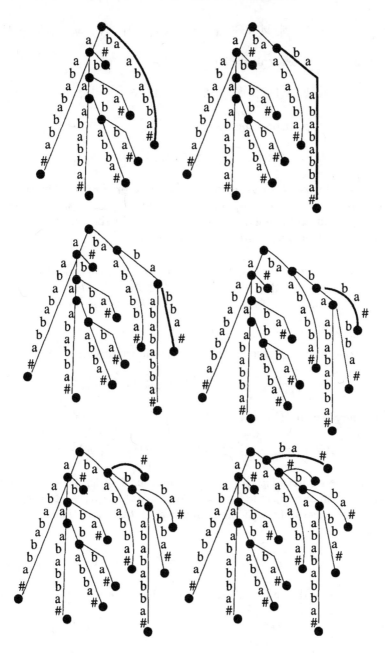

Figure 7.8: The final 6 stages of the algorithm *suffixes-insertion* running on string *babaabababba#*.

$$lcp[rank[1]-1], \ lcp[rank[2]-1], \ lcp[rank[3]-1], \dots, \ lcp[rank[n]-1]$$

in this order. In our example we have:

$$rank \ = \ [8, \ 2, \ 7, \ 1, \ 3, \ 9, \ 4, \ 10, \ 5, \ 12, \ 11, \ 6]$$

and we compute *lcp* for positions

$$7, \ 1, \ 6, \ 0, \ 2, \ 8, \ 3, \ 9, \ 4, \ 11, \ 10, \ 5.$$

Let us denote by $pred(i)$ the index of the suffix preceding p_i in the sorted sequence, so for example $pred(6) = 1$ and $pred(1) = 7$. Then, we consider pairs $(p_{pred(i)}, p_i)$ of adjacent suffixes in the sorted list, in the order of the second element of pairs (the entry with 0 is discarded):

$$(p_{pred(1)}, p_1), \ (p_{pred(2)}, p_2), \ (p_{pred(3)}, p_3), \ \dots, \ (p_{pred(n)}, p_n)$$
$$= \ (p_{i_7}, p_{i_8}), \ (p_{i_1}, p_{i_2}), \ (p_{i_6}, p_{i_7}), \ (p_{i_2}, p_{i_3}), \ (p_{i_8}, p_{i_9}), \ \dots$$
$$= \ (p_3, p_1), \ (p_4, p_2), \ (p_{12}, p_3), \ (p_2, p_5), \ (p_1, p_6), \ \dots$$

because $SufPos \ = \ [4, \ 2, \ 5, \ 7, \ 9, \ 12, \ 3, \ 1, \ 6, \ 8, \ 11, \ 10]$.

The computation is based on the following inequality:

$$lcp[rank[i+1]-1] \geq lcp[rank[i]-1] - 1,$$

which is equivalent to

$$LCPref(p_{pred(i+1)}, p_{i+1}) \geq LCPref(p_{pred(i)}, p_i) - 1.$$

In our example we have $lcp[1] \geq lcs[7] - 1$ and $lcp[6] \geq lcp[1] - 1$. We explain the inequality on our example for $i = 6$. We have that (p_1, p_6) and (p_5, p_7) are pairs of consecutive suffixes in the sorted list. In this case, the inequality can be read as:

$$LCPref(p_5, p_7) \geq LCPref(p_1, p_6) - 1.$$

Observe that p_7 is the suffix p_6 with the first letter deleted. Hence

$$LCPref(p_5, p_7) \geq LCPref(p_2, p_7) \text{ and } LCPref(p_5, p_7) = LCPref(p_1, p_6) - 1.$$

Due to the inequality, when we compute the longest common prefix for the next pair of suffixes we can start from the previous value minus one. The amortized cost is easily seen to be linear.

7.5   The Lowest-Common-Ancestor dictionary

The lowest common ancestor problem (LCA, in short) is to find in constant time, for any two leaves of the suffix tree, the deepest node which is ancestor of both leaves. From the point of view of stringology, more important is the related problem of longest common prefix (LCPREF): for two positions i and j find in constant time $LCPref(p_i, p_j)$. Any solution to the LCA problem gives easily a solution to the LCPREF problem.

Let us remind that $lcp[i]$ is the length of the common prefix of the i-th and $(i + 1)$-th smallest suffix for the lexicographic ordering. Observe that for $i < j$:

$$LCPref(Suffix(i), Suffix(j)) \; = \; \min\{lcp[t] \; : \; t \in [i..j-1]\}.$$

Hence $LCPref$ calculation reduces to the computation of *range minima*. We can compute a *dictionary of range minima*, *DRM*, is a same way as the dictionary of basic factors. Instead of names we compute minima of numbers. Denote $MIN_k[i] \; = \; \min\{lcp[t] \; : \; t \in [i..i+2^k-1]\}$. If we know the tables *MIN* for each k, $1 \le k \le \log n$, then we can compute a minimum over any interval in constant time by taking minima of (possibly overlapping) subintervals which sizes are powers of two. The dictionary of range minima can be computed in $O(n \log n)$ time and needs $O(n \log n)$ space.

The LCA problem can be reduced to the LCPREF problem by precomputing lowest common ancestors for neighboring leaves (in lexicographic order). Let us denote:

$$lca[i] \; = \; LCA(Suffix(i), Suffix(i+1))$$

where *LCA* refers to the suffix tree. We have the following obvious fact.

Lemma 7.3 *If $LCPref(Suffix(i), Suffix(j)) = lcp[t]$ and $t \in [i..j-1]$ then*

$$LCA(Suffix(i), \; Suffix(j)) = lca[t]$$

Hence we have proved in a very simple way the next statement.

Theorem 7.6 *After $O(n \log n)$ preprocessing time we can answer each LCPref and each LCA query in constant time.*

* The LCA problem with linear-time preprocessing

There are many algorithms to improve the time in Theorem 7.6 to linear. The simplest one is an ingenious application of so-called "Four Russians" trick.

Again we deal with the range minima problem. The clue is the reduction of the size of integers. Instead of the table *lcp* we use the table D such that $D[i]$ is the depth of $lca[i] = LCA(Suffix(i), Suffix(i+1))$ in the suffix tree. The depth is the number of edges from the root to a given node. Now Lemma 7.3 can be replaced by this one.

Lemma 7.4 *If* $\min\{D[k] : k \in [i..j-1]\} = D[t]$ *and* $t \in [i..j-1]$ *then*

$$LCA(Suffix(i), Suffix(j)) = lca[t].$$

Hence the LCA problem is reduced to range minima computations on the table D.

Example. For the example string of the preceding section we have:

$$D = [1, 3, 4, 2, 1, 0, 2, 4, 3, 2, 1].$$

The main trick now is to convert this table to the sequence of numbers differing by +1 or -1 and detect small segments that repeat. We insert between any two adjacent elements the sequence of consecutive integers (in ascending or descending order). For example the table $D = [1, 3, 4, 2, 1, 0, 2, 4, 3, 2, 1]$ is converted into

$$D' = [1, 2, 3, 4, 3, 2, 1, 0, 1, 2, 3, 4, 3, 2, 1].$$

It is easy to see that if D corresponds to a suffix tree then the length of the table at most doubles.

Now we can assume that adjacent entries differ by one only. We partition the table D' into blocks of length $p = \frac{\log n}{2}$. For example assume that $p = 3$, then the table $[1, 2, 3, 4, 3, 2, 1, 0, 1, 2, 3, 4, 3, 2, 1]$ is partitioned into the blocks

$$[1, 2, 3], [4, 3, 2], [1, 0, 1], [2, 3, 4], [3, 2, 1].$$

We compute minima for each block and get the compressed table $[1, 2, 0, 2, 1]$. The range minima problem for this table can be done in $O(\frac{n}{\log n} \cdot \log n)$, *i.e.*, $O(n)$ time and space, by the algorithm from Theorem 7.6. It is enough to precompute a range minima data structure for each block: for a given block the queries are only for positions inside the block. We can scale the blocks by subtracting the first element from each number. For example, this gives only three different blocks:

$$[0, 1, 2], [0, -1, -2], [0, -1, 0], [0, 1, 2], [0, -1, -2].$$

Each block corresponds to a sequence of $p-1$ increments or decrements by $+1$ or -1. There are only 2^{p-1} such sequences, which is $O(\sqrt{n})$. Hence there are only $O(\sqrt{n})$ different blocks. We apply the algorithm of Theorem 7.6 to each of them. In fact a naive algorithm is sufficient since the number of blocks is very small. The blocks with the same structure share the same data structure to support range minima queries inside them.

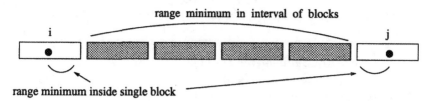

Figure 7.9: A range minimum query is implemented as three queries, two related to a single block, and one related to the compressed table of blocks.

If we have to find the minimum in the interval $[i..j]$ then we check first in which blocks are i and j. Next we compute the minimum in the interval of complete blocks and minima inside single blocks in which are the positions i and j. This takes constant time. The range minimum query is split into two queries related to single blocks and one query related to interval of blocks, which can be answered with the dictionary for the compressed table D' (see Figure 7.9). In this way we have proved the following theorem.

Theorem 7.7 *After $O(n)$ preprocessing time we can answer each LCPref and each LCA query in constant time.*

7.6 Suffix-Merge-Sort

It is easy to sort all suffixes of a string in $O(n\log n)$ time using its dictionary of basic factors or its suffix tree. McCreight and Ukkonen algorithms work in $O(n\log|A|)$ time for the alphabet A.   Assume that the alphabet is a set of integers in the range $[1\dots n]$, so the alphabet can be sorted in linear time. Farach gave the first algorithm which works for this case *really* in linear time, the constant coefficient does not depend on the size of the alphabet. However the improvement from $O(n\log|A|)$ to $O(n)$ is at the cost of a very complex construction.

Basically, as we have seen in Section 7.4, the suffix-tree construction reduces to the computation of the sorted sequence $\mathcal{R}$ = *Suffix-Merge-Sort*(x), for a given text x. Denote by $\mathcal{R}_{even}$ and $\mathcal{R}_{odd}$ the sorted sequences of suffixes starting at odd positions $1, 3, 5, \ldots$ and at even positions, respectively. Let us sort the set of all pairs of adjacent symbols (integers) in the text x and let $rank(a, b)$ be the rank of a pair (a, b). Assume the text x has an even length, $x = a_1 a_2 \ldots a_{2m}$ (if necessary some dummy symbol can be appended). Denote

$$compress(a_1 a_2 \ldots a_{2m}) =$$
$$(rank(a_1, a_2), rank(a_3, a_4), rank(a_5, a_6), \ldots, rank(a_{2m-1}, a_{2m})).$$

Algorithm *Suffix-Merge-Sort*(x);
{ a version of Farach algorithm }
if $|x| = 1$ **then return** single suffix of x
else begin
    Step 1: $\mathcal{R}_{odd} := $ *Suffix-Merge-Sort*$(compress(x))$;
    Step 2: compute $\mathcal{R}_{even}$ using $\mathcal{R}_{odd}$;
    Step 3: construct *Odd/Even-Oracle*;
    Step 4: **return** *MERGE*(S_{odd}, S_{even});
end

For example $compress(2, 3, 1, 2, 3, 2, 1, 2, 3, 4) = (2, 1, 2, 1, 3)$. We can identify $\mathcal{R}_{odd}$ with the sorted sequence of suffixes of $compress(x)$. Each suffix starting at an odd position can be treated as a sequence of pairs of adjacent symbols $a_{2i-1} a_{2i}$ encoded by their rank. At the same time we identify the sorted sequence of suffixes with the sequence of their starting positions in x, which is essentially the same as the table *SufPos*.

We explain each of the steps on the example string $x = babaababababba\#$.

Step 1 The only combination of letters on odd/even neighboring positions are ab and ba, $\#\#$ (we add $\#$'s at the end to have them in pairs). If we encode ab by $\tilde{1}$ and ba by $\tilde{2}$ we obtain:

$$compress(babaababababba\# = \tilde{2}\,\tilde{2}\,\tilde{1}\,\tilde{1}\,\tilde{1}\,\tilde{2}\,\#$$

that is twice shorter than x. The sorted sequence of its suffixes (after re-scaling the positions from $compress(x)$ to x) gives the following sorted sequence of suffixes starting at odd positions in x:

$$\mathcal{R}_{odd} = [5, 7, 9, 3, 1, 11].$$

Step 2 The sorted sequence of odd suffixes is used to give a single integer name to each of them: its rank in the sorted sequence. The suffix at odd position i can be now identified with the pair $(x[i], rank[i+1])$. The sequence of consecutive even suffixes becomes now:

$$(S_2, S_4, S_6, S_8, S_{10}, S_{12}) = (a, \tilde{2}), \ (a, \tilde{1}), \ (b, \tilde{1}), \ (b, \tilde{1}), \ (b, \tilde{2}), \ (a, \#).$$

We use *Sort-Rename* to sort such pairs, the sorting order being equivalent to the sorted order of even suffixes. This gives:

$$\mathcal{R}_{even} = [4, \ 2, \ 12, \ 6, \ 8, \ 10].$$

Step 3 *Odd/Even-Oracle* should give in constant time $LCPref(S_i, S_j)$ for any pair of odd/even positions. This is the most complicated step and is shortly described later.

Step 4 Using *Odd/Even-Oracle* we can compare lexicographically any pair of odd/even suffixes in constant time. We are doing a standard linear-time merging of two sorted sequences, with each comparison in constant time. The output is:

$$Suffix\text{-}Merge\text{-}Sort(x) = \mathcal{R} = [4, \ 2, \ 5, \ 7, \ 9, \ 12, \ 3, \ 1, \ 6, \ 8, \ 11, \ 10].$$

Complexity analysis. If we do not count the complexity of the recursive call then the total complexity of all steps is $O(n)$. Observe that size of *compress(x)* is twice smaller than that of x. Hence the total complexity $T(n)$ can be estimated using the recurrence:

$$T(n) \ \leq \ T(\tfrac{n}{2}) + O(n).$$

Consequently $T(n) = O(n)$.

* Linear-time construction of *Odd/Even-Oracle*

We sketch only the main ideas behind this quite involved construction. In Section 7.4 we showed that suffix trees can be extremely easily constructed in $O(n)$ time if the sorted sequence $\mathcal{R}$ of suffixes is given, the linear time does not dependent on the size of the alphabet. Denote the corresponding procedure by $TreeConstruct(\mathcal{R})$. The sequence $\mathcal{R}$ is given by the table *SufPos*. A rough structure of the *Odd/Even-Oracle* construction is as follows:

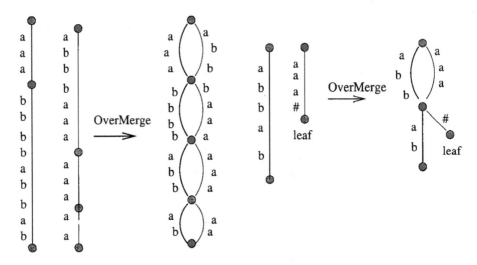

Figure 7.10: Two types of overmerging corresponding paths in T_{odd} and T_{even}.

Algorithm *Construct-Oracle*;
   $T_{odd} := TreeConstruct(\mathcal{R}_{odd})$;
   $T_{even} := TreeConstruct(\mathcal{R}_{even})$;
   $T := OverMerge(T_{odd}, T_{even})$;
   compute Lowest-Common-Ancestor dictionary of T;
   compute d-link for each odd-even node $v \in T$;
   for each $v \in T$ **do**
      $\hat{L}(v) :=$ depth of v in the tree of d-links;
   return *Odd/Even-Oracle* dictionary $=$
      tree T, table $\hat{L}$, and LCA-dictionary of T;

The *LCPref* queries for odd i and even j are answered by *Odd/Even-Oracle* as follows:

$$LCPref(S_i, S_j) = \hat{L}(LCA_T(S_i, S_j)).$$

Each suffix S_i corresponds to a leaf of T. In the trees T_{odd}, T_{even} identify the node i with the suffix S_i starting at position i.

Description of operation *OverMerge*. The paths from T_{odd} and T_{even} are merged, an operation of merging two paths is schematically illustrated in

Figure 7.10. If two edges start with the same letter and are of different lengths
then the longer one is broken and a new node is created. If the shorter edge
ends with a leaf then we break it just before the last letter, to guarantee that
the leafs of T_{odd} and T_{even} correspond to different leaves in the tree T. If the
edges start with the same letter and have the same length then we create a
double edge, consisting of both edges with end-points identified. Figure 7.11
shows the result of overmerging the trees T_{odd} and T_{even} for an example string.

Description of table of d-links. Considering $Overmerge(T_{odd}, T_{even})$, we
say that a node v in it is an odd/even node iff it has two descendant leaves
corresponding to an odd i and to an even leaf j such that $LCA(i, j) = v$. For
each odd/even node v define:

$$d(v) \;=\; LCA(i+1, j+1).$$

Denote by $\hat{L}(v)$ the depth of v in the tree of d-links; it is the smallest k such
that $d^k(v) = root$. Some d-links are shown in Figure 7.11. We refer the reader
to [FFM 00] for the proof of the following lemma.

Lemma 7.5 [Correctness of Odd/Even Oracle] *If* $T = OverMerge(T_{odd}, T_{even})$,
i is odd and j is even, then

$$LCPref(S_i, \; S_j) \;=\; \hat{L}(LCA_T(S_i, \; S_j).$$

The result of the lemma enables to find $LCPref(S_i, S_j)$ in constant time, if
the tree T, the table $\hat{L}$ and the LCA dictionary for T are precomputed (which
takes altogether linear time). Although it is a linear-time construction on the
whole, the overmerging operation is rather overloaded.

Bibliographic notes

The algorithm *KMR* is from Karp, Miller, and Rosenberg [KMR 72]. Another
approach is presented in [Cr 81]. It is based on a modification of Hopcroft's
partitioning algorithm [Ho 71] (see [AHU 74]), and the algorithm computes
vectors $Name_r$ for all values of r with the same complexity.

The notion of *suffix array* was invented by Manber and Myers [MM 90].
They use a different approach to the construction of *suffix arrays* than our
exposition (which refers to the dictionary of basic factors). The worst-case
complexities of the two approaches are identical, though the average complexity
of the original solution proposed by Manber and Myers is lower. A kind of
implementation of *suffix arrays* is considered by Gonnet and Baeza-Yates in

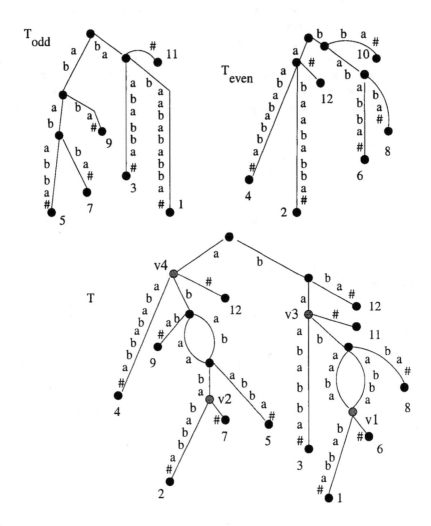

Figure 7.11:   Illustration of algorithm *Construct-Oracle* on the word *babaabababba#*: T_{odd}, T_{even}, and $T = OverMerge(T_{odd}, T_{even})$. We have: $d[v1] = v2$, $d[v2] = v3$, $d[v3] = v4$, $d[v4] = root$. Hence $\hat{L}[v1] = 4$, since the *d*-links depth of $v1$ is 4, and $LCPref(1,6) = 4$. We have also: $LCPref(S_1, S_6) = LCPref(S_2, S_7) + 1 = LCPref(S_3, S_8) + 2 = LCPref(S_4, S_9) + 3 = LCPref(S_5, S_{10}) + 4 = 0 + 4$.

[GB 91], and called *Pat trees*. The interesting linear-time computation of table *lcp* is from Kasai et al. [K-P 01].

The first linear-time alphabet-independent algorithm for integer alphabets was given by Farach [Fa97]. It is an important theoretical milestone in suffix tree constructions. The algorithm is still too complicated. We have shown a much simpler algorithm, with the same complexity, assuming that the sorted sequence of suffixes is already given. Suffix-Merge-Sort is a version of the algorithm in [Fa 97] and [FFM 00].

Chapter 8

Symmetries and repetitions in texts

This chapter presents algorithms dealing with regularities in texts. By regularity we mean a similarity between one segment and some other factors of the text. Such a similarity can be at least of two kinds: one segment is an exact copy of the other, or it may be a symmetric copy of the other.

Algorithmically, an interesting exercise is to detect situations in which similar factors are consecutive in the text. For exact copies we get repetitions of the same factor in the form xx (squares). In the case of symmetric copies we have words in the form xx^R, called even palindromes. Odd palindromes are also interesting regularities (words of the form xax^R, where x is a non-empty word). The compositions of very regular words are in some sense also regular: a palstar is a composition of palindromes. Algorithms of this chapter are aimed at discovering these types of regularities in text.

8.1   Searching for symmetric words

We start with a decision problem that consists of verifying if a given text has a prefix that is a palindrome; such palindromes are called prefix palindromes. There is a very simple linear-time algorithm used to search for a prefix palindrome:

- compute the failure function *Bord* of *text*&*text*^{R} (word of length $2n+1$),

- then, *text* has a prefix palindrome iff $Bord(2n + 1) \neq 0$.

However, this algorithm has two drawbacks: it is not an on-line algorithm, and moreover, we could expect to have an algorithm that computes the smallest prefix palindrome in time proportional to the length of this palindrome (if a prefix palindrome exists). Later in the chapter (when testing palstars) it will be seen why we impose such requirements.

For the time being, we restrict ourselves to even palindromes. In order to improve on the above solution, we proceed in a similar way as in the derivation of *KMP* algorithm: the efficient algorithm is derived from an initial *brute-force* algorithm by examining its invariants. The time complexity of the algorithm below is quadratic (in the worst case). A simple instance of a worst text for the algorithm is $text = ab^n$. The key to the improvement is to make an appropriate use of the information gathered by the algorithm. This information is expressed by invariant

$$w(i, j) : text[i - j + 1 . . i] = text[i + 1 . . i + j].$$

Algorithm *brute-force*;
{ looking for prefix even palindromes }
    $i := 1$;
    while $i \leq \lfloor n/2 \rfloor$ **do begin**
        { check if $text[1 . . 2i]$ is a palindrome }
        $j := 0$;
        while $j < i$ **and** $text[i - j] = text[i + 1 + j]$ **do** $j := j + 1$;
        if $j = i$ **then return** true;
        { $inv(i, j)$: $text[i - j] \neq text[i + j + 1]$ }
        $i := i + 1$;
    end;
    return false;

The maximum value of j satisfying $w(i, j)$ for a given position i is called the radius of the palindrome centered at i, and denoted by $Rad[i]$. Hence, algorithm *brute-force* computes values of $Rad[i]$ but does not profit from their computation. The information is wasted, because at the next iteration, the value of j is reset to 0. As an alternative, we try to make use of all possible information, and, for that purpose, the computed values of $Rad[i]$ are stored in a table for further use. The computation of prefix palindromes easily reduces to the computation of table Rad. Hence, we convert the decision version of algorithm *brute-force* into its optimized version computing Rad. For simplicity, assume that the text starts with a special symbol.

The key to the improvement is not only a mere recording of table Rad, but also a surprising combinatorial fact about symmetries in words. Suppose that

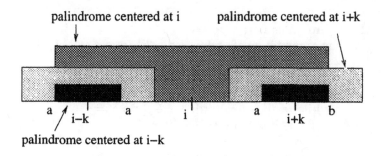

Figure 8.1: Case (b) of proof of Lemma 8.1.

we have already computed the sequence:

$$Rad[1], Rad[2], \ldots, Rad[i].$$

It happens that we can sometimes compute many new entries of table Rad without comparing any symbols. The following fact enables us to do so.

Lemma 8.1 *If the integer k is such that $1 \leq k \leq Rad[i]$ and $Rad[i - k] \neq Rad[i] - k$, then $Rad[i + k] = \min(Rad[i - k], Rad[i] - k)$.*

Proof. Two cases are considered:

Case (a): $Rad[i - k] < Rad[i] - k$.
    The palindrome of radius $Rad[i - k]$ centered at $i - k$ is completely contained inside the longest palindrome centered at i. Position $i - k$ is symmetrical to $i + k$ with respect to i. Hence, by symmetry with respect to position i, the longest palindrome centered at $i + k$ has the same radius as the palindrome centered at $i - k$. This implies the conclusion in this case: $Rad[i + k] = Rad[i - k]$.

Case (b): $Rad[i - k] > Rad[i] - k$.
    The situation is illustrated in Figure 8.1. The maximal palindromes centered at $i - k$, i and $i + k$ are presented. Symbols a and b are distinct due to the maximality of the palindrome centered at i. Hence, $Rad[i + k] = Rad[i] - k$ in this case.

Collecting the results of the two cases, we get $Rad[i + k] = \min(Rad[i - k], Rad[i] - k)$. □

```
Algorithm Manacher;
{ on-line computation of prefix even palindromes and of table Rad;
   text starts with a unique left end-marker }
   i := 2; Rad[1] := 0; j := 0; { j = Rad[i] }
   while i ≤ ⌊n/2⌋ do begin
      while text[i − j] = text[i + 1 + j] do j := j + 1;
      if j = i then write(i); Rad[i] := j;
      k := 1;
      while Rad[i − k] ≠ Rad[i] − k do begin
         Rad[i + k] := min(Rad[i − k], Rad[i] − k); k := k + 1;
      end;
      { inv(i, j):  text[i − j] ≠ text[i + j + 1] }
      j := max(j − k, 0);
      i := i + k;
   end
```

During one stage of the algorithm that computes prefix palindromes, we can update $Rad[i + k]$ for all consecutive positions $k = 1, 2, \ldots$ such that $Rad[i - k] \neq Rad[i] - k$. If the last such k is k', we can then later consider the next value of i as $i + k'$, and start the next stage. The strategy applied here is similar to shifts applied in string-matching algorithms in which values result from a precise consideration on invariants. We obtain the algorithm *Manacher* for on-line recognition of prefix even palindromes and computation of the table of radii.

The solution presented for the computation of prefix even palindromes adjusts easily to the table of radii of odd palindromes. The same holds true for longest palindromes occurring in the text. Several other problems can be solved in a straightforward way using the table *Rad*.

Theorem 8.1 *The longest symmetrical factor and the longest (or shortest) prefix palindrome of a text can be computed in linear time. If text has a prefix palindrome, and if s is the length of the smallest prefix palindrome, then s can be computed in time $O(s)$.*

8.2 Compositions of symmetric words

We now consider another question regarding regularities in texts. Let P^* be the set of words that are compositions of even palindromes, and let PAL^* denote the set of words composed of any type of palindromes (even or odd).

Recall that one-letter words are not palindromes according to our definition (their symmetry is too trivial to be considered as an "interesting" feature).

Our aim now is to test whether a word is an *even palstar*, (a member of P^*) or a *palstar* (a member of PAL^*). We begin with the simpler case of even palstars.

Let *first(text)* be a function which value is the first position i in text such that $text[1..i]$ is an *even* palindrome; it is zero if there is no such prefix palindrome. The following algorithm tests even palstars in a natural way. It finds the first prefix even palindrome and cuts it. Afterward, the same process is repeated as often as possible. If we are left with an empty string, then the initial word is an even palstar.

Theorem 8.2 *Even palstars can be tested on-line in linear time.*

Proof. It is obvious that the complexity is linear, even on-line. In fact, *Manacher* algorithm computes the function *first* on-line. An easy modification of the algorithm gives an on-line algorithm operating within the same complexity.

However, a more difficult issue is the correctness of function *PSTAR*. Suppose that *text* is an even palstar. It then seems reasonable to expect that its decomposition into even palindromes does not necessarily start with the shortest prefix even palindrome.

```
function PSTAR(text); { is text an even palstar ? }
    s := 0;
    while s < n do begin
        if first(text[s + 1..n]) = 0 then return false;
        s := s + first(text[s + 1..n]);
    end;
    return true;
```

Fortunately, and perhaps surprisingly, it so happens that we have always a good decomposition (starting with the smallest prefix palindrome) if *text* is an even palstar. So the greedy strategy of function *PSTAR* is correct. To prove this fact, we need some notation related to decompositions. It is defined only for texts that are even palindromes. Let

$$parse(text) = \min\{s : text[1..s] \in P \text{ and } text[s + 1..n] \in P^*\}.$$

Now the correctness of the algorithm results directly from the following fact.

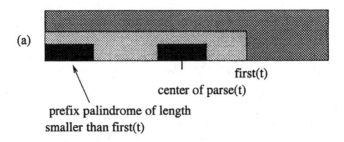

(a)

first(t)

center of parse(t)

prefix palindrome of length
smaller than first(t)

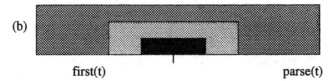

(b)

first(t)   parse(t)

text[1 .. parse(t)] decomposes into at least 2 palindromes

Figure 8.2: The proof (by contradiction) of Theorem 8.2.

Claim. If *text* is a non-empty palstar, then *parse*(*text*) = *first*(*text*).

Proof of the claim. It follows from the definitions that *first*(*text*) ≤ *parse*(*text*), hence, it is sufficient to prove that the reverse inequality holds. The proof is by contradiction. Assume that *text* is an even palstar such that *first*(*text*) < *parse*(*text*). Consider two cases

Case (a): *parse*(*text*)/2 < *first*(*text*) < *parse*(*text*),

Case (b): 2 ≤ *first*(*text*) ≤ *parse*(*text*)/2.

The proof of the claim according to these cases is given in Figure 8.2. This ends the proof of the theorem.   □

If we try to extend the previous algorithm to all palstars, we are led to consider functions *first*1 and *parse*1, analogue to *first* and *parse*, defined respectively as follows:

$$parse1(text) = \min\{s : text[1..s] \in PAL \text{ and } text[s+1..n] \in PAL^*\},$$
$$first1(text) = \min\{s : text[1..s] \in PAL\}.$$

Unfortunately, when *text* is a palstar, the equation *parse*1(*text*) = *first*1(*text*) is not always true. A counter-example is the text *text* = *bbabb*. We have

$parse1(text) = 5$ and $first1(text) = 2$. If $text = abbabba$, then $parse1(text) = 7$
and $first1(text) = 4$. The third case is for $text = aabab$: we have $parse1(text) =$
$first1(text)$. Observe that for the first text, we have the equality $parse1(text) =$
$2 \cdot first1(text) + 1$; for the second text, we have $parse1(text) = 2.first1(text) - 1$;
and for the third text, we have $parse1(text) = first1(text)$. It happens that it
is a general rule that only these cases are possible.

Lemma 8.2
Let x be a palstar, then $parse1(x) \in \{first1(x), 2.first1(x) - 1, 2.first1(x) + 1\}$.

Proof. The proof is similar to the proof of the preceding lemma. In fact,
the two special cases $(2.first1(text) \pm 1)$ are caused by the irregularity implied at critical points by considering odd and even palindromes together. Let
$f = first1(text)$, and $p = parse1(text)$. The proof of the impossibility of the
situation $f < p < 2.f - 1$ is essentially presented in the case (a) of Figure 8.2.
The proof of the impossibility of the two other cases, $p = 2.f$ and $p > 2.f + 1$,
is similar. □

Theorem 8.3 *Palstar membership can be tested in linear time.*

Proof. Assume that we have computed the tables

$$F[i] = first1(text[i + 1..n]) \text{ and } PAL[i] = (text[i + 1..n] \text{ is a palindrome}).$$

Then it is easy to see that the algorithm *PALSTAR* recognizes palstars in
linear time.

Assuming the correctness of function *PALSTAR*, in order to prove that it
works in linear time, it is sufficient to show how to compute tables F and PAL
in linear time. The computation of PAL is trivial if the table Rad is known.
This latter can be computed in linear time.

The computation of table F is more difficult. For simplicity we restrict ourselves to odd palindromes, and compute the table

$$F1[s] = \min\{s : text[1..s] \text{ is an odd palindrome or } s = 0\}.$$

Assume Rad is the table of radii of odd palindromes: the radius of a palindrome
of size $2k + 1$ is k. We say that j is in the range of an odd palindrome centered
at i iff $i - j + 1 \leq Rad[i]$ (see Figure 8.3).

A stack of positions is used. It is convenient to have some special position at
the bottom. Initially the first position 1 is pushed onto the stack, and i is set
to 2. One stage begins with the situation depicted in Figure 8.3. All x's are
positions put in the stack (entries of $F1$ waiting for their values). The whole
algorithm is:

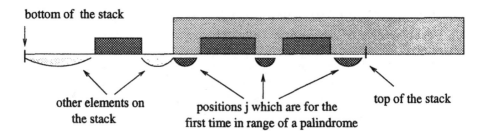

Figure 8.3: Stage (i) in the proof of Theorem 8.3: **while** the top position j is in the range of the palindrome centered at i **do begin** $pop(stack); F1[j] := 2(j - i) + 1$ **end**; $push(i)$.

> **for** $i := 2$ **to** n **do** stage(i);
> **for** all remaining elements j on the stack **do**
>   { j is not in a range of any palindrome } $F1[j] := 0$

The treatment of even palindromes is similar. In that case, $F[i]$ is computed as the minimum value of even or odd palindromes starting at position i. This completes the proof.   □

> **function** $PALSTAR(text)$: Boolean; { palstar recognition }
> $palstar[n] :=$ true; { the empty word is a palstar }
>   **for** $i := n - 1$ **down to** 0 **do begin**
> $f := F[i]$;
>   **if** $f = 0$ **then** $palstar[i] :=$ true
>   **else if** $PAL[i]$ **then** $palstar[i] :=$ true
>   **else** $palstar[i] := (palstar[i + f]$ **or** $palstar[i + 2f - 1]$
>   **or** $palstar[i + 2f + 1])$
>   **end**;
>   **return** $palstar[0]$;

It is perhaps surprising that testing whether a text is a composition of a fixed number of palindromes is more difficult than   testing for palstars. Recall that here P denotes the set of even palindromes. It is very easy to recognize compositions of exactly two words from P. The word $text$ is such a composition iff, for some internal position i, $text[1..i]$ and $text[i + 1..n]$ are even palindromes. This can be checked in linear time if table Rad is already computed. But this approach does not directly produce a linear-time algorithm for P^3. Fortunately, there is another combinatorial property of texts that is useful for that case. Its proof is omitted.

Lemma 8.3 *If $x \in P^2$, then there is a decomposition of x into words $x[1..s]$ and $x[s..n]$ such that both are members of P, and that either the first word is the longest prefix palindrome of x, or the second one is the longest suffix palindrome of x.*

One can compute the tables of all the longest palindromes starting or ending at positions on the word by a linear-time algorithm very similar to that computing table F. Assume now that we have these tables, and also the table *Rad*. One can then check if any suffix or prefix of *text* is a member of P^2 in constant time thanks to Lemma 8.3 (only two positions are to be checked using preprocessed tables).

Now we are ready to recognize elements of P^3 in linear time. For each position i we just check in constant time if $text[1..i] \in P^2$ and $text[i+1..n] \in P$. Similarly, we can test elements of P^4. For each position i we check in constant time if $text[1..i] \in P^2$ and $text[i+1..n] \in P^2$. We have just sketched the proof of the following statement.

Theorem 8.4 *The compositions of two, three, and four palindromes can be tested in linear time.*

As far as we know, there is presently no algorithm to test compositions of exactly five palindromes in linear time. We have defined palindromes as non trivial symmetric words (words of size at least two). One can say that for a fixed k, palindromes of a size smaller than k are uninteresting. This leads to the definition of PAL_k as palindromes of size at least k. Generalized palstars (compositions of words from PAL_k) can also be defined. For a fixed k, there are linear-time algorithms for such palstars. The structure of algorithms, and the combinatorics of such palindromes and palstars are analogous to what is presented in the section.

8.3 Searching for square factors

It is a non trivial problem to find a square factor in linear time, that is, a non-empty factor in the form xx in a text. A naive algorithm gives cubic bound on the number of steps. A simple application of failure functions gives a quadratic algorithm. For that purpose, we can compute failure tables $Bord_i$ for each suffix $text[i..n]$ of the text. Then, there is a square starting at position

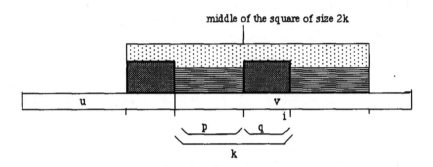

Figure 8.4: A square xx of size $2k$ occurs in uv. The suffix of $v[1..k]$ of size q is also a suffix of u; the prefix of $v[k+1..|v|]$ of size p is a prefix of v. We have $PREF[k] + SUF_u[k] \geq k$.

i in the text iff $Bord_i[j] \geq j/2$, for some $1 \leq j \leq n - i + 1$. Since each failure table is computed in linear time, the whole algorithm takes quadratic time.

We develop an $O(n \log n)$ algorithm that tests the squarefreeness of texts, and afterward design a linear-time algorithm (for fixed alphabets). The first method is based on a divide-and-conquer approach. The main feature of both algorithms is a fast implementation of the Boolean function $test(u, v)$ that tests whether the word uv contains a square, for two squarefree words u and v. Then, if uv contains a square, it must begin in u and ends in v. Thus, the operation $test$ is a composition of two smaller Boolean functions: $righttest$ and $lefttest$. The first one searches for a square for which the center is on v, while the second searches for a square for which the center is on u.

We now describe how $righttest$ works on words u and v. We use two auxiliary tables related to string matching. The first table $PREF$ is defined on the word v as in Section 3.2. For a position k on v, $PREF[k]$ is the size of longest prefix of v occurring at position k (it is a prefix of $v[k+1..|v|]$). The second table is called SUF. The value $SUF_u[k]$ (k is still a position on v) is the size of longest suffix of $v[1..k]$ that is also a suffix of u. Table SUF_u is a generalization of table S discussed in Chapter 3. These tables can be computed in linear time with respect to $|v|$. With the two tables, the existence of a square in uv centered on v reduces to a simple test on each position of v, as shown by Figure 8.4.

Lemma 8.4 *The Boolean value $righttest(u, v)$ can be computed in $O(|v|)$ time (independently of $|u|$).*

Proof. The computation of tables $PREF$ and SUF_u takes $O(|v|)$ time. It is clear (see Figure 8.4) that there exists a square centered on v iff, for some position k, $PREF[k] + SUF_u[k] \geq k$. All these tests take again $O(|v|)$ time. □

The dual version of Lemma 8.4 states that the value $lefttest(u, v)$ can be computed in $O(|u|)$ time, which gives the next result.

Corollary 8.1 *The Boolean value $test(u,v)$ can be computed in $O(|u| + |v|)$ time.*

Proof. Compute $righttest(u, v)$ and $lefttest(u, v)$. The value $test(u, v)$ is the Boolean value: $righttest(u, v)$ or $lefttest(u, v)$.
The computation takes $O(|v|)$ (Lemma 8.4), and $O(|u|)$ time (by symmetry from Lemma 8.4), respectively. The result follows. □

---

function $SQUARE(text)$: Boolean;
  { checks if $text$ contains a square, $n = |text|$ }
   if $n > 1$ **then begin**
      if $SQUARE(text[1..\lfloor n/2 \rfloor])$ **then return** true;
      if $SQUARE(text[\lfloor n/2 \rfloor + 1..n])$ **then return** true;
      if $test(text[1..\lfloor n/2 \rfloor], text[\lfloor n/2 \rfloor + 1..n])$ **then return** true;
   end;
   return false; { if value true not yet returned }

Theorem 8.5 *Algorithm SQUARE tests whether a text of length n contains a square in time $O(n \log n)$.*

Proof. The algorithm has $O(n \log n)$ time complexity, because $test$ can be computed in linear time. The complexity can be estimated using a divide-and-conquer recurrence. □

The algorithm $SQUARE$ inherently runs in $O(n \log n)$ time. This is due to the divide-and-conquer strategy for the problem. But, it can be shown that the algorithm extends to the detection of all squares in a text. And this shows that the algorithm becomes optimal, because some texts may contain exactly $O(n \log n)$ squares. For example this happens for Fibonacci words.

We show that the question of testing the squarefreeness of a word can be answered in linear time on a fixed alphabet. This contrasts with the previous problem. The strategy uses again a kind of divide-and-conquer, but with an unbalanced nature. It is based on a special factorization of texts. Our interest

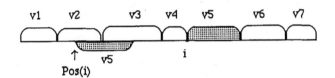

Figure 8.5: Efficient factorization of the source, v_5 is the longest factor occurring before position i.

is mainly theoretical because of a rather large overhead in memory usage. The algorithm, or more exactly the factorization defined for the purpose, is related to data compression methods based on the elimination of repetitions of factors. The algorithm shows another profound application of data structures developed in Chapters 4–6 for storing all factors of a text.

We first define the f-factorization of $text$ (the f stands for factors) (see Figure 8.5). It is a sequence of non-empty words $(v_1, v_2, \ldots, v_m)$, where

- $v_1 = text[1]$, and

- for $k > 1$, v_k is defined as follows. If $|v_1 v_2 \ldots v_{k-1}| = i$ then v_k is the longest prefix u of $text[i+1..n]$ that occurs at least twice in $text[1..i]u$. If there is no such u then $v_k = text[i+1]$. We denote by $pos(v_k)$ the smallest position $l < i$ such that an occurrence of v_k starts at l. If there is no such position then set $pos(v_k) = 0$.

---

function $linear\text{-}SQUARE(text)$: Boolean;
{ checks if $text$ contains a square, $n = |text|$ }
    compute the f-factorization $(v_1, v_2, \ldots, v_m)$ of $text$;
    for $k := 1$ **to** m **do**
        if at least one of the conditions of Lemma 8.5 holds **then**
            return true;
    return false; { if true has not been returned yet }

The f-factorization of a $text$, and the computation of values $pos(v_k)$ can be realized in linear time with the directed acyclic word graph $G = DAWG(text)$ or with the suffix tree $T(text)$. Indeed, the factorization is computed while the DAWG is built. The overall procedure has the same asymptotic linear bound as just building the DAWG. This leads to the final result of the section, consequence of the following observation which proof is left as an exercise.

Lemma 8.5 *Let* $(v_1, v_2, \ldots, v_m)$ *be the f-factorization of text. Then, text contains a square iff for some k at least one of the following conditions holds:*

1. $|v_1 v_2 \ldots v_{k-1}| \le pos(v_k) + |v_k| < |v_1 v_2 \ldots v_k|$,

2. *lefttest*(v_{k-1}, v_k) *or righttest*(v_{k-1}, v_k),

3. *righttest*$(v_1 v_2 \ldots v_{k-2}, v_{k-1} v_k)$.

Theorem 8.6 *Function linear-SQUARE tests whether a text of length n contains a square in time $O(n)$ (on a fixed alphabet), with $O(n)$ additional memory space.*

Proof. The computation of *righttest*$(v_1 v_2 \ldots v_{k-2}, v_{k-1} v_k)$ is the key point: it can be executed in $O(|v_{k-1} v_k|)$ time. Thus the total time is proportional to the sum of length of all v_k's; hence it is linear. This completes the proof. □

Bibliographic notes

The algorithm for prefix palindromes is from Manacher [Ma 75]. The material in the rest of the section and in Section 8.2 is mainly from Galil and Seiferas [GS 78]. The reader can refer to [GS 78] for the proof of Lemma 8.3, and an on-line linear-time algorithm for *PALSTAR*.

The first $O(n \log n)$ algorithm for searching a square is by Main and Lorentz [ML 79] (see also [ML 84]). The linear-time algorithm of Section 8.3 is from Crochemore [Cr 83] (see also [Cr 86]). The f-factorization introduced here is a variation of the Ziv-Lempel decomposition [ZL 77] used for text compression and presented in Chapter 10.

A different method for achieving linear time square testing has been proposed by Main and Lorentz [ML 85].

Using suffix trees, Stoye and Gusfield show how to compute repetitions efficiently in [SG 98].

Chapter 9

Constant-space searchings

The most intriguing algorithms are those efficient simultaneously with respect to two measures of complexity. The linear-time string matching with constant space is a question of this type. There are several different time-space efficient algorithms for this problem, all rely on simple combinatorics of periodicities in texts.

9.1 Constant-space matching for easy patterns

We start with a family of patterns (self-maximal strings) that are very *easy* to be searched for in texts in constant space and $O(n)$ time.

Denote by $MaxSuf(w)$ the lexicographically maximal suffix of the word w. The word w is said to be *self-maximal* if $MaxSuf(pat) = pat$.

Maximal suffixes of words and *self-maximal* words play an important role in the computation of periods for several reasons (recall that $period(x)$ is the smallest period of x, see Chapter 1s):

(1) If *pat* is periodic then $period(MaxSuf(pat)) = period(pat)$.

(2) If *pat* is *self-maximal* then each of its prefixes also is.

(3) If *pat* is *self-maximal* then $period(pat)$ can be trivially computed by the following function:

We consider the function *Naive-Period* that correctly computes the period of a word if this word is self-maximal.

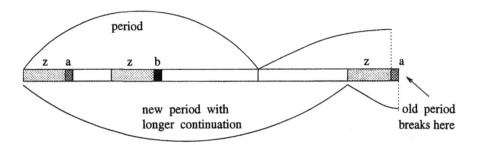

Figure 9.1: Assume, in the algorithm *Naive-Period*, that $pat[i - period(i-1)] \neq pat[i]$. Let $a = pat[i]$, $b = pat[i - period]$. Since uz is a prefix of *pat* which is self-maximal, we have and $a < b$. If $Period(i) < i$ then, due to the two periodicities, zb is a proper subword of $pat[1..i - 1]$ that is lexicographically greater than *pat*. This contradicts the self-maximality of *pat*. Hence $Period(i) = i$.

---

function *Naive-Period(j)*;
  { computes the period of self-maximal *pat* }
    period := 1;
    for $i := 2$ **to** j **do**
      if $pat[i] \neq pat[i - period]$ **then** *period* := i;
    return *period*;

Example. The function *Naive-Period* usually gives incorrect output for non-self-maximal words. For example, consider the string

$$pat = (aba)^6 = abaabaabaabaabaaba.$$

The consecutive values of *period* computed by the function for all positions are:

a	b	a	a	b	a	a	b	a	a	b	a	a	b	a	a	b	a
1	2	2	4	5	5	7	8	8	10	11	11	13	14	14	16	17	17

Hence *Naive-Period*(18) = 17, for $pat = (aba)^6$, while $period(pat) = 3$.

Lemma 9.1 *Assume pat is a self-maximal string. The function Naive-Period computes correctly the exact period of pat, as well as of each prefix of pat.*

Proof. An informal justification is given in Figure 9.1.　　　　　　□

We modify *MP* algorithm (Morris-Pratt algorithm) that uses one additional table related to the pattern and a constant number of additional registers. We consider how to get rid of the table. Recall that the basic table needed in *MP* algorithm is the table $MP\_Shift[j] = j - Bord[j]$. Recall also that $Bord[j]$ is precisely the length of the largest proper border of $pat[1..j]$. Therefore

$$MP\_Shift[j] = Period(j)$$

for $j > 0$, where $Period(j) = period(pat[1..j])$. The *MP* algorithm can be re-written by changing *MP_Shift* to $Period(j)$, assuming $Period(0) = 0$.

In addition, for self-maximal patterns we can embed the computation of *Naive-Period*(j) directly into *MP* algorithm. $Period(j)$ is computed here "on-the-fly" in an especially simple way. Doing so, we get the algorithm called *SpecialCase-MP*. The next lemma follows directly from the correctness of both algorithm *Naive-Period*(j) and algorithm *MP*.

Lemma 9.2 *If the pattern is self-maximal then we can find all its occurrences in $O(1)$ space and linear time with the algorithm SpecialCase-MP.*

Algorithm *SpecialCase-MP*;
　　$i := 0;\ j := 0;\ period := 1;$
　　while $i \leq n - m$ **do begin**
　　　　while $j < m$ and $pat[j+1] = text[i+j+1]$ **do begin**
　　　　　　$j = j + 1;$
　　　　　　if $j > period$ **and** $pat[j] \neq pat[j - period]$ **then**
　　　　　　　　$period := j;$
　　　　end;
　　　　MATCH: if $j = m$ **then return** match at i;
　　　　$i := i + period;$
　　　　if $j \geq 2 \cdot period$ **then** $j := j - period;$
　　　　else begin $j := 0;\ period := 1;$ **end;**
　　end;
　　return no match;

9.2　MaxSuffix-Matching

In this section we develop the first time-space optimal string-matching algorithm, which is a natural extension of Morris-Pratt algorithm and which as-

$$pat \;=\; Fib_7 \;=\; abaababaabaababaababa \;=\; \overbrace{abaababaabaa}^{u} \; \overbrace{babaababa}^{v}$$

$$text \;=\; abaabaa \, \overbrace{babaababa}^{v} \, baa \, \overbrace{\underline{babaababa}}^{v} \, \underline{abaa} \, \overbrace{babaababa}^{v} \, ba$$

Figure 9.2: The decomposition of the 7-th Fibonacci word and history of *MaxSuffix-Matching* on an example text. Only for the third occurrence of v we check if there is an occurrence of u (underlined segment) to the left of this occurrence of v, because gaps between previous occurrences of v are too small. The first occurrence is too close to the beginning of *text*.

sumes that the alphabet is ordered. We assume that *pat* and *text* are *read-only* arrays, and we count the additional memory space as the number of integer registers (each in the range $[0..n]$). We assume that the decomposition $pat = u \cdot v$, where $v = MaxSuf(pat)$, is known. Algorithm *SpecialCase-MP* can be used to report each occurrence of v. We extend this algorithm to general patterns by testing in a *naive* way occurrences of u to the left of v. But we do not need to test for u to the left of v for each occurrence of v due to the following fact (see Figure 9.2).

Lemma 9.3 [Key-Lemma] *Assume that an occurrence of v starts at position i on text. Then, no occurrence of uv starts at any of the positions $i - |u| + 1, i - |u| + 2, \ldots, i$ on text.*

Proof. This is because the maximal suffix v of *pat* can start at only one position on *pat*.   □

The algorithm that implement the remark of Lemma 9.3 is presented below. It is followed by a less informal description.

Algorithm *Informal-MaxSuffix-Matching*;
     Let $pat = uv$, where $v = MaxSuf(pat)$;
     Search for v with algorithm *SpecialCase-MP*;
     for each occurrence of v at i on *text* **do**
         Let *prev* be the previous occurrence of v;
         if $i - prev > |u|$ **then**
             if u occurs to the left of v **then** report a *match*;
             { occurrence of u is tested in a *naive* way }

Theorem 9.1 *Algorithm MaxSuffix-Matching makes at most n extra symbol comparisons in addition to the total cost of searching for v.*

Proof. Additional naive tests for occurrences of u are related to disjoint segments of *text*, due to Lemma 9.3. Then, altogether this takes at most n extra symbol comparisons. □

Algorithm *MaxSuffix-Matching*;
   $i := 0$; $j := 0$; *period* := 1; *prev* := 0;
   while $i \leq n - |v|$ **do begin**
      while $j < |v|$ **and** $v[j+1] = text[i+j+1]$ **do begin**
         $j = j+1$;
         if $j > period$ **and** $v[j] \neq v[j - period]$ **then** *period* := j;
      end;
      { **MATCH OF** v: } **if** $j = |v|$ **then**
         if $i - prev > |u|$ **and** $u = text[i - |u| + 1..i]$ **then begin**
            report a match at $i - |u|$;
            prev := i;
         end;
      $i := i + period$;
      if $j \geq 2 \cdot period$ **then** $j := j - period$
      else begin $j := 0$; *period* := 1; **end**;
   end

9.3 Computation of maximal suffixes

We can convert the function *Naive-Period* into a computation of the length of the longest self-maximal prefix of a given text x. The algorithm simultaneously computes its shortest period. The correctness is straightforward from the previous discussion.

---

function *Longest-Self-Maximal-Prefix*(x);
   period := 1;
   for $i := 2$ **to** $|x|$ **do**
      if $x[i] < x[i - period]$ **then** *period* := i
      else if $x[i] > x[i - period]$ **then**
         return $(i - 1, period)$;
   return $(|x|, period)$ { there was no return earlier };

We use the algorithm *Longest-Self-Maximal-Prefix* as a key component in the computation of the maximal suffix of the whole text. The function *MaxSuf-and-Period0* returns the starting position of the maximal suffix of its input and the period of the maximal suffix.

> **function** *MaxSuf-and-Period0* (x);
> $j := 1$;
>   **repeat**
> $(i, period) := $ *Longest-Self-Maximal-Prefix* $(x[j . . n])$;
>   **if** $i = n$ **then return** $(j, period)$
>   **else** $j := j + i - (i$ **mod** *period*$)$;
>   **forever** { a return breaks the loop }

Example. Let $x = abcbcbacbcbacbc$. Then, for the usual ordering of letters,

$$MaxSuf(x) = cbcbacbcbacbc,$$

which is also $(cbcba)^2 cbc$. The maximal suffix of xa is:

$$cbcbacbcbacbca = MaxSuf(x)a,$$

which is a border-free word. The maximal suffix of xb is

$$cbcbacbcbacbcb = MaxSuf(x)b.$$

This word has the same period as $MaxSuf(x)$. Finally, the maximal suffix of xc is cc.

We can rewrite the function *MaxSuf-and-Period0* as a self-contained function which does not use calls to other functions. The code is shorter but looks more tricky. It is impressive that such a short code describes a rather complex computation of maximum suffix together with its period in linear time and constant additional space. The function   mod   can even be removed at the expense of a slightly more complex structure of the algorithm, without affecting the linear-time complexity.

Lemma 9.4 *The algorithm MaxSuf-and-Period makes less than $2.|x|$ letter comparisons.*

Proof. The value of expression $s + j + k$ is increased by at least one unit after each symbol comparison. The result then follows from inequalities $2 \leq s + j + k \leq 2.|x| + 1$. □

```
function Maxsuf-and-Period(x);
    s := 1; i := 2; p := 1;
    while (i ≤ n) do begin
        r := (i - s)  mod p;
        if (x[i] = x[s + r]) then i := i + 1
        else if (x[i] < x[s + r]) then begin
            i := i + 1; p := i - s;
        end else begin
            s := i - r; i := s + 1; p := 1;
        end;
    end;
    return (s, p);
    { x[s..n] = MaxSuf(x), p = period(MaxSuf(x)) }
```

9.4 Matching patterns with short maximal suffixes

Maximal suffixes have many unexpected properties. We use the following simple property of maximal suffixes which proof is left to the reader (see Figure 9.3).

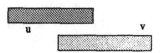

Figure 9.3: An overlap between u and v is impossible when v is the maximal suffix of uv.

Lemma 9.5 *Let* $v = MaxSuf(x)$ *and* $x = uv$. *Then* u *and* v *cannot overlap: no non-empty suffix of* u *can be a prefix of* v.

In this section we consider a special case of patterns x with short maximal suffixes. They satisfy:

$$|MaxSuf(x)| \leq |x|/2.$$

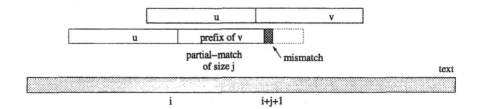

Figure 9.4: Illustration of the proof of the Short Prefixes Lemma: impossible situation. Any next occurrence of v can start only to the right of position $i + j + 1$ because an overlap between u and v is impossible.

Example. Let $x = abababaaababaababababababab$.
Then x decomposes into $uMaxSuf(x)$ with

$$u = abababaaababaaa \text{ and } MaxSuf(x) = babababababab.$$

Lemma 9.6 [Short Prefixes Lemma] *Assume the pattern pat has a short maximal suffix. Let pat = uv be the decomposition of the pattern ($v = MaxSuf(pat)$). Assume that we align the pattern pat starting at position i, scan part v and find the first mismatch at the $(j + 1)$-th position on v. Then we can skip safely the pattern by $j + 1$ positions without missing any occurrence of pat in the text.*

Proof. Assume that *pat* is aligned with *text* in such a way that $text[i]$ is aligned with the last letter of u. If we shift *pat* and align the segment v with the text at a position between i and $i + j + 1$, then u and v will overlap on the previously matched part of v, see Figure 9.4. But due to Lemma 9.5, u and v cannot have a nonempty overlap. A contradiction that proves the result.   □

We can now modify the *MaxSuffix-Matching* algorithm by inserting directly the shift rule implied by the above lemma. It gives the instruction:

if $j < |v|$ **then** $Shift(j) = j + 1$ **else** $Shift(j) = period(v)$.

The resulting algorithm is called *Two-Way Pattern-Matching*.

Theorem 9.2 *Assume $|MaxSuf(pat)| \leq |pat|/2$. Algorithm Two-Way Pattern-Matching finds all occurrences of the pattern in $O(1)$ space using at most $2n$ symbol comparisons, if the maximal suffix $v = MaxSuf(pat)$ is precomputed.*

Proof. Due to large shifts we never test a position of v against the same position of the text twice. Tests for u are done on disjoint segments of the text. This gives at most $n + n$ comparisons.   □

Algorithm *Two-way Pattern-Matching*;
{ Simplified version of Crochemore-Perrin algorithm }
{ *period* = the period of v }
    $i := |u|$; $j := 0$; *prev* := 0;
   while $i \leq n - |v|$ **do begin**
     while $j < |v|$ **and** $v[j+1] = text[i+j+1]$ **do** $j = j+1$;
     { **MATCH OF** v: } **if** $j = |v|$ **then begin**
       if $i - prev > |u|$ **and** $u = text[i - |u| + 1 .. i]$ **then**
         report a match at $i - |u|$;
       prev := i; $i := i + period$; $j := |v| - period$;
     end else
     { **MISMATCH of** v: } $i := i + j + 1$;
   end

9.5 Two-way matching and magic decomposition

For general patterns the algorithm of the preceding section would be incorrect. The algorithm is based on the interaction between the two parts u and v of the pattern. The nightmare scenario for this algorithm is the case of self-maximal patterns because then u is empty and there is no useful interaction at all. Fortunately, there is a surprising way to get around this problem. Consider two inverse orderings of the alphabet, compute the maximal suffixes with respect to these orderings, and keep the shorter of them. It does not have to be short in the sense of the preceding section, but it is short enough to guarantee the correctness of the two-way pattern matching.

We consider two alphabetical orderings on words. The first one, denoted by $\leq$, is induced by a given ordering $\leq$ on the alphabet. The second ordering on words, called the *reverse ordering* and denoted by $\subseteq$, is obtained by reversing the order $\leq$ on the alphabet A.

Magic decomposition. Let x be a non-empty word on A. Let $x = u_1 v_1 = u_2 v_2$, where v_1 (resp. v_2) is the alphabetically maximal suffix of x according to the ordering $\leq$ (resp. $\subseteq$). If $|v_1| \leq |v_2|$, then (u_1, v_1) is the *magic decomposition* of x. Otherwise (u_2, v_2) is the magic decomposition of x.

Let (u, v) be the magic decomposition of *pat* and assume that the pattern is nontrivial, which means here that it contains at least two distinct letters. In this case certainly u is a non-empty word.

```
                                    *    *    *    *    *    *    *
                     a    a    b    a    a    b    b    b    a    b    a    a
               a    a    b    a    a    b    b
     a    a    b    b    a    b    a    a    b    a    a    b    a    a    b    b    a    b    a    a
```

Figure 9.5: Searching pattern $x = u \cdot v = aabbab \cdot aabaabbabaa$. Text is the bottom line. Since $|u| = 6$, we start searching for v at position 7; the mismatch is at position 7 on v, so the shift has length 7 and we start testing v at position 14, aligning v at positions indicated by $*$, despite the fact that we missed the match of v at position 10. But this occurrence cannot be preceded by u, so no occurrence of uv is missed.

Example. Let $x = aabbabaabaabbabaa$. Then

$$magic\text{-}decomposition(x) \;=\; u \cdot v \;=\; aabbab \cdot aabaabbabaa.$$

Figure 9.5 gives an example of the use of this decomposition.

Theorem 9.3 *Assume the pattern contains at least two distinct symbols. The algorithm Two-Way Pattern-Matching is correct if (u, v) is the magic decomposition of the pattern.*

Proof. We only need to show that Lemma 9.6 works also in the case where (u, v) is the magic decomposition of *pat*, with u nonempty. The following fact is obvious and we omit its proof.

Claim 1. For any words x and y, inequalities $x \leq y$ and $x \subseteq y$ are equivalent to x is a prefix of y.

Claim 2. Assume u occurs in v: $v = zuy$ for some z, y. Then $|zu|$ is a period of the whole pattern.
Proof of the claim. Let $w = zu$ (Figure 9.6).
Without loss of generality we may assume that v is the maximal suffix for the ordering $\leq$. Let $pat = u'v'$ where v' is the maximal suffix for $\subseteq$. Let u'' be the non-empty word such that $u = u'u''$ (see Figure 9.7). Recall that hypothesis $|v| < |v'|$ implies that u' is a proper prefix of u. Since $u''y$ is a suffix of x, by the definition of v', we get $u''y \subseteq v' = u''v$, hence $y \subseteq v$. By the definition of v, we also have $y < v$. By Claim 1, these two inequalities imply that y is a prefix of v. Hence, $pat = uzuy$, where y is a prefix of zuy. This shows that $|zu|$ is a period of v and of the whole pattern, which completes the proof of the claim.

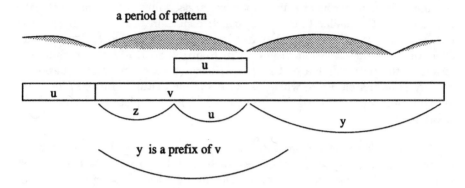

Figure 9.6: When (u, v) is the magic decomposition of the pattern, if u occurs fully inside v then $|zu|$ is a global period of the pattern.

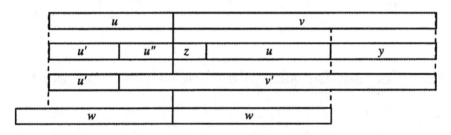

Figure 9.7: Illustration of the proof that y is a prefix of zuy.

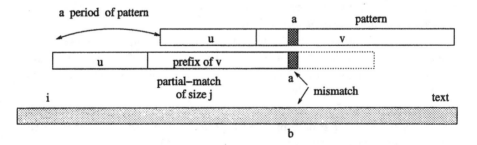

Figure 9.8: Any second occurrence of u inside the pattern determines a period of it. Then, the mismatch recurs.

Assume now that we have a mismatch when scanning v and that we make a shift of length smaller than $j + 1$, which results in placing u "inside" the matched part of u. Observe that overlap is impossible due to Lemma 9.5. Then we have the situation displayed in Figure 9.8. However the shift is a period of the pattern, due to Claim 2. Hence the same mismatch happens, so there is no match of the whole pattern starting at this position. Therefore any shift smaller than $j + 1$ can be skipped. This complete the proof of the theorem. □

For a position i on the word x a *local period* at i is any positive integer *per* such that $x[i - j] = x[i - j + per]$ for each j, $0 \leq j \leq per - 1$, for which both sides of the equation are defined. Denote by $LocalPeriod(x, i)$ the size of a minimal local period at i. We say that i is a *critical factorization point* if $period(x) = LocalPeriod(x, i)$. As a side effect of the proofs of the last theorem we have the following corollary.

Theorem 9.4 *Assume x contains at least two distinct symbols and let (u, v) be its magic decomposition. Then $|u|$ is a critical factorization point of x. A critical factorization point can be computed in constant space and linear time.*

9.6 Sequential sampling for unordered alphabets

When the only access to the input data is by testing equality of two symbols, we cannot use the previous algorithms *MaxSuffix-matching* nor *Two-Way Pattern-Matching* since they are based on maximal suffixes.

We describe shortly the searching phases of two algorithms that work optimally in this situation: *Sequential-Sampling* in the section and the Galil-Seiferas algorithm in the next section.

The searching phases of both algorithms are simple, but the preprocessing phases are not. Another disadvantage is that the cases of periodic and non-periodic patterns should be considered separately. This was not necessary for maximal-suffix-based algorithms.

Recall that a text x is *periodic* if $period(x) \leq |x|/2$. We consider only the case when pat is not periodic but has a periodic prefix of pat. Let j be the length of the longest periodic prefix of pat. Define the set

$$S = \{ p, q \}, \text{ where } p = j + 1 - period(pat[1..j]), q = j + 1.$$

The set S is called the *sample* of pat. The positions p and q are the first (from the left) *witnesses against* the period *period*(*pat*[1..*j*] of *pat*. This means that $q - p = period(pat[1..q-1])$ and $pat[p] \neq pat[q]$. Let us introduce the predicate:

$$MatchSample(i, S) = (text[i + p] = pat[p] \text{ and } text[i + q] = pat[q]).$$

Observation. If *MatchSample*(*i*, *S*) holds then no occurrence of the pattern starts at any position in $i + 1, i + 2, \ldots, i + p$ on *text*. The observation implies that if the pattern matches the text at the positions of the sample, then the next *safe shift* has length at least *p*. For example if *pat* = *aaaaaaab* then $S = \{7, 8\}$. In this case if *MatchSample*(*i*, *S*) then the next shift has length at least 8.

Algorithm *Sequential-Sampling*; { Searching phase only }
{ the case where *pat* has a sample $S = \{p, q\}$ }
    i := 0;
    while $i \leq n - m$ **do begin**
        if $pat[p] \neq text[i + p]$ **or** $pat[q] \neq text[i + q]$ **then** $i := i + 1$
        else begin
            j := 0;
            while $j < m$ **and** $pat[j + 1] = text[i + j + 1]$ **do** $j := j + 1$;
        MATCH: if $j = m$ **then** report a match at *i*;
        MISMATCH: if $j < q - 1$ **then** $i := i + p$ **else** $i := i + \lceil \frac{j+1}{2} \rceil$;
        end
    end

Remark. In the algorithm, we assume that, when computing $j := \max\{k : text[i + 1..i + k] = pat[1..k] \text{ or } k = 0\}$, symbols $text[i + p]$ and $text[i + q]$ are not tested again, since we have already tested them when computing *MatchSample*(*i*, *S*). We say that a word x is highly periodic if $|x| \geq period(x)/3$.

Theorem 9.5 *Assume pat is not highly periodic but has a highly periodic prefix. Then there exist a sample of pat for which the algorithm Sequential-Sampling performs report the occurrences of pat in text. The algorithm makes at most 2n symbol comparisons and uses a constant additional space.*

Proof. Let $pat[1..q - 1]$ be the longest periodic prefix of *pat*, *per* be the period of $pat[1..q - 1]$ and $S = \{q - per, q\}$. Negative tests on letters are amortized by immediate shifts, i.e. two comparisons are amortized by a shift

of length one. In case of a positive match of the sample S, we start to test
the full match of $pat[1..q]$, omitting earlier recognized symbols of the sample.
Symbols at positions in S do not belong to the period of the prefix $pat[1..q-1]$,
so if a mismatch between $text$ and the prefix is found we can make a shift of
length $s = p$. Hence the total work is not greater than $q-1$ and $p \geq \frac{q-1}{2}$ (prefix
$pat[1..q-1]$ is periodic) so $q \leq 2 \cdot s$, and we get the proper amortization. □

9.7   Galil-Seiferas algorithm

The algorithm of Galil and Seiferas (GS algorithm in short) can be regarded
as another space-efficient implementation of MP algorithm. In the context
of MP algorithm, the idea behind saving on space is to avoid storing all the
periods of prefixes of the pattern. Only small periods are memorized, where
"small" is relative to the length of the prefix. Approximations of other periods
are computed when necessary during the search phase.

Highly repeating prefixes (hrp's). A primitive word w is called a *highly
repeating prefix* of x (a hrp of x, in short) if w^3 is a prefix of x. Recall that a
word w is said to be *primitive* if it is not a (proper) power of any other word.
Primitive words are non-empty, so are hrp's.

GS-decomposition.   We say that a word is *GS-good* if it has at most one
hrp. The decomposition (u, v) of x is said to be a *GS-decomposition* if

$$|u| < 2.period(v) \text{ and } v \text{ has at most one hrp.}$$

The basic idea behind GS algorithm is to scan the pattern from a position where a GS-good word starts. Fortunately, such a position always exists
because words satisfy a remarkable combinatorial property stated in the next
theorem, and which technical proof is omitted (we refer the interested reader
to [CR 94]).

Theorem 9.6 [GS-decomposition theorem] *Any non-empty word x can be factorized into a GS-decomposition in $O(1)$ space and linear time.*

Consider a GS-good word v as a pattern. Let p be the length of the only
hrp of v and let r be the length of the longest prefix of v having period p.
Then we can search for pattern v using a version of MP algorithm with the
shift function computed in constant space and constant time as follows:

(1) $Shift(j) = p$ if $j \in [2p..r]$,

(2) *Shift*$(j) = \lceil \frac{i}{3} \rceil$, otherwise.

In the first case we reset j to $j - p$, in the second case we reset j to 0. The cost of comparisons is amortized by shifts, and this proves the following.

Lemma 9.7 *If v is a GS-good word, then we can search for v in constant space and linear time with MP algorithm modified by the shift function Shift.*

If the pattern *pat* is not GS-good, we consider its GS-decomposition (u, v). Searching for the whole pattern is done by the previous search for its part v, together with naive tests for the prefix part u. Since $|u| < 2.period(v)$, comparisons for the latter tests are no more than $|text|$. This gives the following informal description of Galil-Seiferas algorithm. Observe that it is quite similar to MaxSuffix-Matching, where we have a part v (self-maximal word) which search is simple and the other part u that is tested naively.

Algorithm *GS*; { informal description }
{ search for *pat* in *text* }
    $(u, v) :=$ GS-decomposition of *pat*;
    find all occurrences of v in *text* using
    MP algorithm with the modified shift function;
    for each position q of an occurrence of v in *text* **do**
        if u ends at q **then** report a match at position $q - |u|$;
        { occurrences of u are tested in a naive way }

9.8 Cyclic equality of words

A *rotation* of a word u of length n is any word of the form $u[k + 1 . . n]u[1 . . k]$, denoted by $u^{(k)}$ (note that $u^{(0)} = u^{(n)} = u$). Let u, w be two words of the same length n. They are said to be *cyclic-equivalent* if $u^{(i)} = w^{(j)}$ for some i, j. If words u and w are written as circles, they are cyclic-equivalent if the circles coincide after appropriate rotations.

There are several linear-time algorithms for testing the cyclic-equivalence of two words. The simplest one is to apply any string matching algorithm to pattern *pat* $= u$ and *text* $= ww$ because words u and w are cyclic-equivalent iff *pat* occurs in *text*.

Another algorithm is to find maximal suffixes of uu and ww and check if they are identical on prefixes of size n. We have chosen this problem because there is simpler interesting algorithm, working in linear time and constant space simultaneously, which deserves presentation.

Define $D(u)$ and $D(w)$ as:

$$D(u) = \{k : 1 \le k \le n \text{ and } u^{(k)} > w^{(j)} \text{ for some } j\},$$
$$D(w) = \{k : 1 \le k \le n \text{ and } w^{(k)} > u^{(j)} \text{ for some } j\}.$$

We use the following simple fact:

if $D(u) = [1..n]$ or $D(w) = [1..n]$, then words u, w are not cyclic equivalent.

Now the correctness of the algorithm follows from preserving the invariant:

$$D(w) \supseteq [1..i] \text{ and } D(u) \supseteq [1..j].$$

Algorithm *Cyclic-Equivalence*(u, w)
{ checks cyclic equality of u and w of common length n }
    $x := uu$; $y := ww$;
    $i := 0$; $j := 0$;
    while $(i < n)$ **and** $(j < n)$ **do begin**
        $k := 1$;
        while $x[i + k] = y[j + k]$ **do** $k := k + 1$;
        if $k > n$ **then return** true;
        if $x[i + k] > y[i + k]$ **then** $i := i + k$ **else** $j := j + k$;
        { invariant }
    end;
    return false;

The number of symbol comparisons is linear. The largest number of comparisons is for words in the forms $u = 111\ldots1201$ and $w = 1111\ldots120$.

Bibliographic notes

The first time-space optimal string-matching algorithm is from Galil and Seiferas [GS 83]. The same authors have designed other string-matching algorithms requiring only a small memory space [GS 80], [GS 81]. *MaxSuffix-Matching* is from Rytter [Ry 02].

Two-Way Pattern-Matching is from Crochemore and Perrin [CP 91]. The proof of the magic lemma is based on *critical factorization*. And the proof of the critical factorization theorem can be found in [Lo 83], [CP 91] or [CR 94]. The computation of maximal suffixes is adapted from an algorithm of Duval [Du 83].

Chapter 10

Text compression techniques

The aim of data compression is to provide representations of data in a reduced form. The information carried by data is left unchanged by the compression processes considered in this chapter. There is no loss of information. In other words, the compression processes that we discuss are reversible.

The main interest of data compression is its practical nature. Methods are used both to reduce the memory space required to store files on hard disks or other similar devices, and to accelerate the transmission of data in telecommunications. This feature remains important particularly due to the rapid increase of mass memory, because the amount of data increases accordingly (to store images produced by satellites or medical scanners, for example). The same argument applies to transmission of data, even if the capacity of existing media is constantly improved.

We describe data compression methods based on substitutions. The methods are general, which means that they apply to data about which little information is known. Semantic data compression techniques are not considered. Therefore, compression ratios must be appreciated on that condition. Standard methods usually save about 50% memory space.

Data compression methods attempt to eliminate redundancies, regularities, and repetitions in order to compress the data. It is not surprising then that algorithms have features in commons with others described in preceding chapters.

After Section 10.1 on elementary notions about the compression problem, we consider the classical Huffman statistical coding (Sections 10.2 and 10.3).

It is implemented by the UNIX (system V) command "pack." It admits an adaptive version well suited for telecommunications, and implemented by the "compact" command of UNIX (BSD 4.2). Section 10.4 deals with the general problem of factor encoding, and contains the efficient Ziv-Lempel algorithm.

The "compress" command of UNIX (BSD 4.2) is based on a variant of this latter algorithm.

10.1   Substitutions

The input of a data compression algorithm is a text. It is denoted by s, for *source*. It should be considered as a string on the alphabet $\{0, 1\}$. The output of the algorithm is also a word of $\{0, 1\}^*$ denoted by c, for *encoded* text. Data compression methods based on substitutions are often described with the aid of an intermediate alphabet A on which the source s translates into a text *text*. A method is then defined by mean of two morphisms g and h from A^* to $\{0, 1\}^*$. The text *text* is an inverse image of s by the morphism g, which means that its letters are coded with bits. The encoded text c is the direct image of *text* by the morphism h. The set $\{(g(a), h(a)) : a \in A\}$ is called the *dictionary* of the coding method. When the morphism g is known or implicit, the description of the dictionary is given simply by the set $\{a, h(a) : a \in A\}$.

We only consider data compression methods that have no loss of information. This implies that a *decoding function* f exists such that $s = f(c)$. Again, f is often defined through a decoding function h' such that *text* $= h'(c)$, and then f itself is the composition of h' and g. The lossless information constraint implies that the morphism h is one-to-one, and that h' is its inverse morphism. This means that the set $\{h(a) : a \in A\}$ is a uniquely decipherable code.

The pair of morphisms, (g, h), Leads to a classification of data compression methods with substitutions. We get four principal classes according to whether g is uniform (*i.e.*, all images of letters by g are words of the same length) or not, and whether the dictionary is fixed or computed during the compression process. Most elementary methods do not use any dictionary. Strings of a given length are sometimes called *blocks* in this context, while factors of variable lengths are called *variables*. A method is then qualified as *block-to-variable* if the morphism g is uniform, or *variable-to-variable* if neither g nor h are assumed to be uniform.

The efficiency of a compression method that encodes a text s into a text c is measured through a *compression ratio*. It can be $|s|/|c|$, or its inverse $|c|/|s|$. It is sometimes more sensible to compute the amount of space saved by the compression: $(|s| - |c|)/|c|$.

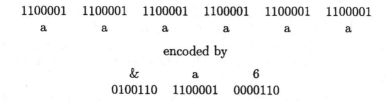

Figure 10.1: Repetition coding (with ASCII code).

	Block-to-Variable	**Variable-to-Variable**
	differential encoding	repetitions encoding
fixed dictionary	statistical encoding (Huffman)	factor encoding
evolutive dictionary	sequential statistical encoding (Faller and Gallager)	Ziv and Lempel's algorithm

The remainder of this section is devoted to the description of two basic methods. They appear on the first line of the previous table *repetitions encoding* and *differential encoding*.

The aim of repetitions encoding is to efficiently encode repetitions. Let *text* be a text on the alphabet A. Let us assume that *text* contains a certain quantity of repetitions of the form $aa\ldots a$ for some character a $(a \in A)$. Within *text*, a sequence of n letters a, can be replaced by $\&an$, where the symbol $\&$ is a new character $(\& \notin A)$. This corresponds to the usual mathematical notation a^n. When all the repetitions of a fixed letter are encoded in such a manner, the letter itself does not need to appear, so that a repetition is encoded by just $\&n$. This is commonly considered for space deletion in text format.

The string $\&an$ that encodes a repetitions of n consecutive occurrences of a is itself encoded on the binary alphabet $\{0, 1\}$. In practice, letters are often represented by their ASCII code. Therefore, the codeword of a letter belongs to $\{0, 1\}^k$ with $k = 7$ or 8. Generally there is no problem in choosing the special character $\&$ (in Figure 10.1, the real ASCII symbol $\&$ is used). Both symbols $\&$ and a appear in the coded text c under their ASCII form. The integer n of the string $\&an$ should also by encoded on the binary alphabet. Note that it is not sufficient to translate n by its binary representation, because we would be unable to localize it at decoding time inside the stream of bits. A simple way to cope with this is to encode n by the string $0^\ell \text{bin}(n)$, where $\text{bin}(n)$ is the binary representation of n, and ℓ is the length of it. This works well because the binary representation of n starts with a 1 (also because $n > 0$). There are even more sophisticated integer representations, but none really suitable for the present situation. Conversely, a simpler solution is to encode n on the same number of bits as letters. If this number is $k = 8$ for example, the translation of any repetitions of length less than 256 has length 3 letters or 24 bits. Thus it is useless to encode a^2 and a^3. A repetition of more than 256 identical letters is then cut into smaller repetitions. This limitation reduces the efficiency of the method.

The second very useful elementary compression technique is differential encoding, also called relative encoding. We explain it using an example. Assume that the source text s is a series of dates

$$1981, 1982, 1980, 1982, 1985, 1984, 1984, 1981, \ldots$$

These dates appear in binary form in s, so at least 11 bits are necessary for each of them. But, the sequence can be represented by the other sequence

$$1981, 1, -2, 2, 3, -1, 0, -3, \ldots$$

assuming that the integer 1 in place of the second date is the difference between the second and the first dates and so on. An integer in the second sequence, except the first one, is the difference between two consecutive dates. The decoding algorithm is straightforward, and processes the sequence from left to right, which is well adapted for texts. Again, the integers of the second sequence appear in binary form in the coded text c. In the example, all but the first can be represented with only 3 bits each. This is how the compression is realized on suitable data.

More generally, differential encoding takes a sequence $(u_1, u_2, \ldots, u_n)$ of data, and represents it by the sequence of differences $(u_1, u_2 - u_1, \ldots, u_n - u_{n-1})$ where $-$ is an appropriate operation.

Differential encoding is commonly used to create archives of successive versions of a text. The initial text is fully memorized on the device. And, for each following version, only the differences from the preceding one are kept on the device. Several variations on this idea are possible. For example, $(u_1, u_2, \ldots, u_n)$ can be translated into $(u_1, u_2 - u_1, \ldots, u_n - u_1)$, considering that the differences are all computed relatively to the first element of the sequence. This element can also change during the process according to some rule.

Very often, several compression methods are combined to realize a whole compression software. A good example of this strategy is given by the application to facsimile (FAX) machines, for which we consider one of the existing protocols. Pages to be transmitted are made of lines, each of 1728 bits. A differential encoding is first applied on consecutive lines. Therefore, if the nth line is

$$0101001010101010010011101001101110110000000 \ldots$$

and the $n + 1^{th}$ line is

0101000010101101110011100101101110100000000...

the following line to be sent is

00000010000001111000000011000000000001000000...

in which a symbol indicates a difference between lines n and $n + 1$. Of course, no compression at all would be achieved if the line were sent explicitly as it is. There are two solutions to encoding the line of differences. The first solution encodes runs of 1 occurring in the line both by their length and their relative position in the line. Therefore, we get the sequence

$$(7, 1), (7, 4), (8, 2), (10, 1), \ldots$$

which representation on the binary alphabet gives the coded text. The second solution makes use of statistical Huffman codes to encode successive runs of $0s$ and runs of $1s$. This type of codes is defined in the next section.

A good compression ratio is generally obtained when the transmitted image contains typeset text. But it is clear that "dense" pages lead to a small compression ratio, and the that best ratios are reached with blank (or black) pages.

10.2 Static Huffman coding

The most common technique for compressing a text is to redefine the codewords associated with the letters of the alphabet. This is achieved by *block-to-variable* compression methods. According to the pair of morphisms (g, h) introduced in Section 10.1, this means that g represents the usual code attributed to the characters (ASCII code, for example). More generally, the source is factorized into blocks of a given length. Once this length is chosen, the method reduces to the computation of a morphism h that minimizes the size of the encoded text $h(text)$. The key to finding h is to consider the frequency of letters, and to encode frequent letters by short codewords. Methods based on this criterion are called *statistical compression methods*.

Computing h requires finding the set $C = \{h(a) : a \in A\}$, which must be a uniquely decipherable code in order to permit the decoding of the compressed text. Moreover, to get an efficient decoding algorithm, C is chosen as an instantaneous code, *i.e.*, a *prefix code*, which means that no word of C is a prefix of another word of C. It is quite surprising that this does not reduce the power of the method. This is due to the fact that any code has an equivalent

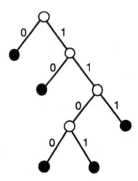

Figure 10.2: The trie of the prefix code $\{0, 10, 1100, 1101, 111\}$. Black nodes correspond to codewords.

prefix code with the same distribution of codeword lengths. The property, stated in the following theorem, is a consequence of what is known as the Kraft-McMillan inequalities related to codeword lengths, which are recalled first.

Kraft's inequality. There is a prefix code with word lengths $\ell_1, \ell_2, \ldots, \ell_k$ on the alphabet $\{0, 1\}$ iff

$$\sum_{i=1}^{k} 2^{-\ell_i} \leq 1 \tag{10.1}$$

McMillan's inequality. There is a uniquely decipherable code with word lengths $\ell_1, \ell_2, \ldots, \ell_k$ on the alphabet $\{0, 1\}$ iff the inequality 10.1 holds.

Theorem 10.1 *A uniquely decipherable code with prescribed word lengths exists iff a prefix code with the same word lengths exists.*

Huffman's method computes the code C according to a given distribution of frequencies of the letters. This method is both optimal and practical. The entire Huffman's compression algorithm proceeds in three steps. In the first step, the numbers of occurrences of letters (blocks) are computed. Let n_a be the number of times letter a occurs in *text*. In the second step, the set $\{n_a : a \in A\}$ is used to compute a prefix code C. Finally, in the third step, the text is encoded with the prefix code C found previously.

Note that the prefix code should be appended to the coded text because the decoder needs it to perform the decompression. It is commonly put inside a *header* (of the compressed file) which contains additional information on the file. Instead of computing the exact numbers of occurrences of letters in the source, a prefix code can be computed on the base of a standard probability distribution of letters. In this situation, only the third step is applied to encode a text, which gives a very simple and fast encoding procedure. Obviously, however, the coding is no longer optimal for a given text.

The core of Huffman algorithm is the computation of a prefix code, over the alphabet $\{0, 1\}$, corresponding to the distribution of letters $\{n_a : a \in A\}$. This part of the algorithm builds the word trie T of the desired prefix code. The prefix property of the code ensures that there is a one-to-one correspondence between codewords and leaves of T (see Figure 10.2). Each codeword (and leaf of the trie) corresponds to some number n_a, and yields the encoding $h(a)$ of the letter a.

The size of the coded text is

$$|h(text)| = \sum_{a \in A} n_a \times |h(a)|.$$

On the trie of the code C the equality translates into

$$|h(text)| = \sum_{a \in A} n_a \times |level(f_a)|$$

where f_a is the leaf of T associated with letter a, and $level(f_a)$ is its distance to the root of T. The problem of computing a prefix code $C = \{h(a) : a \in A\}$ such that $|h(text)|$ is minimum becomes a problem on trees:

- find a minimum weighted binary tree T in which the leaves $\{f_a : a \in A\}$ have initial costs $(n_a : a \in A)$,

where the weight of T, denoted by $W(T)$, is understood as the quantity $\sum n_a \times level(f_a)$. The following algorithm builds a minimum weighted tree in a bottom-up fashion, grouping together two subtrees under a new node. In other words, at each stage the algorithm creates a new node that is made a father of two existing nodes. There are several possible trees of minimum weight, and all trees that can be created by Huffman algorithm are called *Huffman trees*.

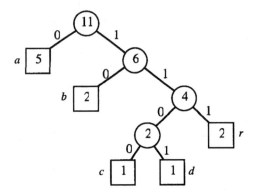

Figure 10.3: Huffman tree for *abracadabra*.

Algorithm *Huffman* { minimum weighted tree construction }
    for a in A **do** create a one-node tree f_a with cost $c(f_a) = n_a$;
    $L :=$ queue of trees f_a in increasing order of costs;
    $N :=$ empty queue; { for internal nodes of the final tree }
    while $|L| + |N| > 1$ **do begin**
        let u and v be the elements of $L \cup N$ with smallest costs;
        { u and v are found from heads of lists }
        delete u and v from (heads of) queues;
        create a tree x with a new node as root,
            and u and v as left and right subtrees;
        $c(x) := c(u) + c(v)$;
        add tree x at the end of queue N;
    end;
    return the remaining tree in $L \cup N$;

Example. Let *text* = *abracadabra*. The number of occurrences of letters are

$$n_a = 5, n_b = 2, n_c = 1, n_d = 1, n_r = 2.$$

The tree of a possible prefix code built by Huffman algorithm is shown in Figure 10.3. The corresponding prefix code is:

$$h(a) = 0, h(b) = 10, h(c) = 1100, h(d) = 1101, h(r) = 111.$$

The coded text is then

$$0101110110011010101110$$

that is a word of length 23. If the initial codewords of letters have length 5, we get the compression ratio $55/23 \approx 2.4$. However, if the prefix code (or its trie) has to be memorized, this additionally takes at least 9 bits, which reduces the compression ratio to $55/32 \approx 1.7$. If the initial coding of letters also has to be stored (with at least 25 bits), the compression ratio is even worse: $55/57 \approx 0.96$. In the latter case, the encoding leads to an expansion of the source text.

As noted in the previous example, the header of the coded text must often contain sufficient information on the coding to allow a later decoding of the text. The information necessary is the prefix code (or information to recompute it) computed by Huffman algorithm, and the initial codewords of letters. Altogether, this takes approximately $2|A| + k|A|$ bits (if k is the length of initial codewords), because the structure of the trie can be encoded with $2|A| - 1$ bits.

Theorem 10.2 *Huffman algorithm produces a minimum weighted tree in time* $O(|A| \log |A|)$.

Proof. The correctness is based on several observations. First, a Huffman tree is a binary complete tree, in the sense that all internal nodes have exactly two sons. Otherwise, the weight of the tree could be decreased by replacing a node having one child by the child itself. Second, there is a Huffman tree T for $\{n_a : a \in A\}$ in which two leaves at the maximum level are siblings, and have minimum weights among all leaves (and all nodes, as well). Possibly exchanging leaves in a Huffman tree gives the conclusion. Third, let x be the father of two leaves f_b and f_c in a Huffman tree T. Consider a weighted tree T' for $\{n_a : a \in A\} - \{n_b, n_c\} + \{n_b + n_c\}$, assuming that x is a leaf of cost $n_b + n_c$. Then,

$$W(T) = W(T') + n_b + n_c.$$

Thus, T is a Huffman tree iff T' is. This is the way the tree is built by the algorithm, joining two trees of minimal weights into a new tree.

The sorting phase of the algorithm takes $O(|A| \log |A|)$ time with any efficient sorting method. The running time of the instructions inside the "while" loop is constant because the minimal weighted nodes in $L \cup N$ are at the beginning of the lists. Therefore, the running time of the "while" loop is proportional to the number of created nodes. Since exactly $|A| - 1$ nodes are created, this takes $O(|A|)$ time. □

The performance of Huffman codes is related to a measure of information of the source text, called the *entropy of the alphabet*. Let p_a be $n_a/|text|$. This quantity can be viewed as the probability that letter a occurs at a given position in the text. This probability is assumed to be independent of the position. Then, the entropy of the alphabet according to p_a's is defined as

$$H(A) = - \sum_{a \in A} p_a \log p_a.$$

The entropy is expressed in bits (log is a base-two logarithm). It is a lower bound of the average length of the codewords $h(a)$,

$$m(A) = \sum_{a \in A} p_a . |h(a)|.$$

Moreover, Huffman codes give the best possible approximation of the entropy (among methods based on a recoding of the alphabet). This is summarized in the following theorem whose proof relies on the Kraft-McMillan inequalities.

Theorem 10.3 *The average length of any uniquely decipherable code on the alphabet A is at least $H(A)$. The average length $m(A)$ of a Huffman code on A satisfies $H(A) \leq m(A) \leq H(A) + 1$.*

The average length of Huffman codes is exactly the entropy $H(A)$ when, for each letter a of A, $p_a = 2^{-|h(a)|}$ (note that the sum of all p_a's is 1). The ratio $H(X)/m(X)$ measures the efficiency of the code. In English, the entropy of letters according to a common probability distribution of letters is close to 4 bits. And the efficiency of a Huffman code is more than 99%. This means that if the English source text is stored as a 8-bit ASCII file, the Huffman compression method is likely to divide its size by two. The Morse code (developed for telegraphic transmissions), which also takes into account probabilities of letters, is not a Huffman code, and has an efficiency close to 66%. This not as good as any Huffman code, but Morse code incorporates redundancies in order to make transmissions safer.

In practice, Huffman codes are built for ASCII letters of the source text, but also for digrams (factors of length 2) occurring in the text instead of letters. In the latter case, the source is factorized into blocks of length 16 bits (or 14 bits). On larger texts, the length of blocks chosen can be higher to capture extra dependencies between consecutive letters, but the size of the alphabet grows exponentially with the length of blocks.

Huffman algorithm generalizes to the construction of prefix codes on alphabet of size m larger than two. The trie of the code is then an almost m-ary tree. Internal nodes have m sons, except maybe one node which is a parent of less than m leaves.

The main default of the entire *Huffman* compression algorithm is that the source text must be read twice: the first time to compute the frequencies of letters, and the second time to encode the text. Only one reading of the text is possible if one uses known average frequencies of letters, but then, the compression is not necessarily optimal on a given text. The next section presents a solution for avoiding two readings of the source text.

There is another statistical encoding that produces a prefix code. It is known as the Shannon-Fano coding. It builds a weighted tree, as Huffman's method does, but the process works top-down. The tree and all its subtrees are balanced according to their costs (sums of occurrences of characters associated with leaves). The result of the method is not necessarily an optimal weighted tree, so its performance is generally within that of Huffman coding.

10.3 Dynamic Huffman coding

The section describes an adaptive version of Huffman's method. With this algorithm, the source text is read only once, which avoids the drawback of the original method. Moreover, the memory space required by the algorithm is proportional to the size of the current trie, that is, to the size of the alphabet. The encoding of letters of the source text is realized while the text is scanned. In some situations, the obtained compression ratio is even better that the corresponding ratio of Huffman's method.

Assume that za (z a word, a a letter) is a prefix of *text*. We consider the alphabet A of letters occurring in z, plus the extra symbol $\#$ that stands for all letters not occurring in z but possibly appearing in *text*. Let us denote by T_z any Huffman tree built on the alphabet $A \cup \{\#\}$ with the following costs:

$$\begin{cases} n_a = & \text{number of occurrences of } a \text{ in } z, \\ n_{\#} = & 0. \end{cases}$$

We denote by $h_z(a)$ the codeword corresponding to a, and determined by the tree T_z.

Note that the tree has only one leaf of null cost, namely, the leaf associate with $\#$.

The encoding of letters proceeds as follows. In the current situation, the prefix z of *text* has already been coded, and we have the tree T_z together with the corresponding encoding function h_z. The next letter a is then encoded by $h_z(a)$ (according the tree T_z). Afterward, the tree is transformed into T_{za}. At decoding time the algorithm reproduces the same tree at the same time. However, the letter a may not occur in z, in which case it cannot be translated as any other letter. In this situation, the letter a is encoded by $h_z(\#)g(a)$, that is, the codeword of the special symbol $\#$ according the tree T_z followed by the initial code of letter a. This step is also reproduced without any problem at decoding time.

The reason why this method is practically effective is due to an efficient procedure updating the tree. The procedure *UPDATE* used in the following algorithm can be implemented in time proportional to the height of the tree. This means that the compression and decompression algorithms work in real time for any fixed alphabet, as it is the case in practice. Moreover, the compression ratio obtained by the method is close to that of *Huffman* compression algorithm.

Algorithm *adaptive-Huffman-coding*
    $T := T_\varepsilon$;
    while not end of *text* **do begin**
        $a :=$ next letter of *text*; { h is implied by T }
        if a is not a new letter **then** write($h(a)$)
        else write($h(\#)g(a)$); { $g(a) =$ initial codeword of a }
        $T := UPDATE(T)$;
    end

The key point of the adaptive compression algorithm is a fast implementation of the procedure *UPDATE*. It is based on a characterization of Huffman trees, known as the siblings property. This property does not hold in general for minimum weighted trees.

Theorem 10.4 [siblings property] *Let T be a complete binary weighted tree (with n leaves) in which leaves have positive costs, and the cost of any internal node is the sum of costs of its children. Then, T is a Huffman tree iff its nodes can be arranged in a sequence $(x_1, x_2, \ldots, x_{2n-1})$ such that:*

1. *the sequence of costs $(c(x_1), c(x_2), \ldots, c(x_{2n-1}))$ is in increasing order, and*

2. *for any i, $1 \le i < n$, the consecutive nodes x_{2i-1} and x_{2i} are siblings (they have the same parent).*

Proof. If T is a tree built by *Huffman* algorithm, the ordering on nodes is simply given by the order in which nodes are deleted from queues during the run of the algorithm.

The "if" part of the proof is by induction on the number n of leaves. Consider the two nodes x_1 and x_2 of the list. It is rather obvious that they are leaves because their costs are strictly positive integers. The leaves x_1 and

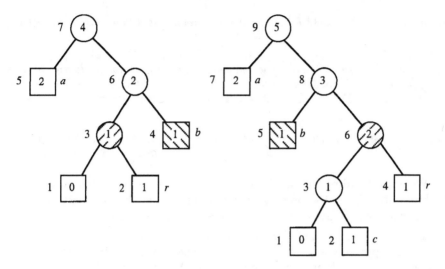

Figure 10.4: Transformation of T_{abra} into T_{abrac}. Marked nodes have been exchanged. Number beside nodes give an ordering satisfying the sibling property.

x_2 can be chosen first by the *Huffman* algorithm because they have minimum costs. Let x be their parent. The rest of the construction is executed as if we had only $n - 1$ leaves, x_1 and x_2 being substituted by x. By induction, the existence of the ordering proves that the tree in which x is a leaf is a Huffman tree. Thus, the initial tree in which x is the parent of x_1 and x_2 is also a Huffman tree (see the proof of Theorem 10.2).   □

   The characterization of Huffman trees by the siblings property remains true if only one leaf has a null cost. During the sequential encoding, the transformation of the current tree T_z into T_{za} starts by incrementing the cost of the leaf x_i that corresponds to a. If point 1 of the siblings property is no longer satisfied, node x_i is exchanged with the node x_j for which j is the greatest integer such that $c(x_j) < c(x_i)$. If necessary, the same operation is repeated on the father of x_i, and so on. The exchange of nodes is, in fact, the exchange of the corresponding subtrees (see Figure 10.4). The tree structure is not affected by exchanges because costs strictly increase from leaves to the root (except maybe in the 3-node tree containing the leaf associated with #), so that a node cannot be exchanged with any of its ancestors. This proves that procedure *UPDATE* can be implemented in time proportional to the height of the tree. Thus, we have the following statement.

Lemma 10.1 *Procedure UPDATE can be implemented to work in time $O(|A|)$.*

Example. Figure 10.5 shows the sequential encoding of *abracadabra*. Letters are assumed to be initially encoded on 5 bits ($a \to 00000$, $b \to 00001$, $c \to 00010, \ldots, z \to 11010$). The entire translation of *abracadabra* is:

00000	000001	0010001	0	10000010	0	110000011	0	110	110	0
a	*b*	*r*	*a*	*c*	*a*	*d*	*a*	*b*	*r*	*a*

We get a word of length 45. These 45 bits are to be compared with the 57 bits obtained by the original *Huffman* algorithm. For this example, the dynamic algorithm gives a better compression ratio, say $55/45 \approx 1.22$.

The precise analysis of the adaptive *Huffman* compression algorithm has led to an improved compression algorithm. The key for the improvement is to choose a specific ordering for the nodes of the Huffman tree. Indeed, one may note that in the ordering given by the siblings property, two nodes of same cost are exchangeable. The improvement is based on a ordering that corresponds to a width-first tree-traversal of the tree from leaves to the root. Moreover, at each level in the tree, nodes are ordered from left to right, and leaves of a given cost precede internal nodes having the same cost. The algorithm derived from this idea is efficient for texts of just a few thousand characters. The encoding of larger texts save almost one bit per character compared with Huffman algorithm.

10.4   Factor encoding

Data compression methods with substitutions gain their full power when the substitution applies to variable-length factors rather than blocks. The substitution is defined by a dictionary:

$$D = \{(f, c) : f \in F, c \in C\},$$

where F is a (finite) set of factors of the source text s, and C is the set of their corresponding codewords. The set F acts as the alphabet A of Section 10.1. The source text is a concatenation of elements of F.

Example. Let *text* be a text composed of ordinary ASCII characters encoded on 8 bits. For C, one may choose the 8-bit words, if any, that correspond to no letter of *text*. Then, F can be a set of factors occurring frequently inside the text *text*. Replacing factors of F in *text* by letters of C compresses the

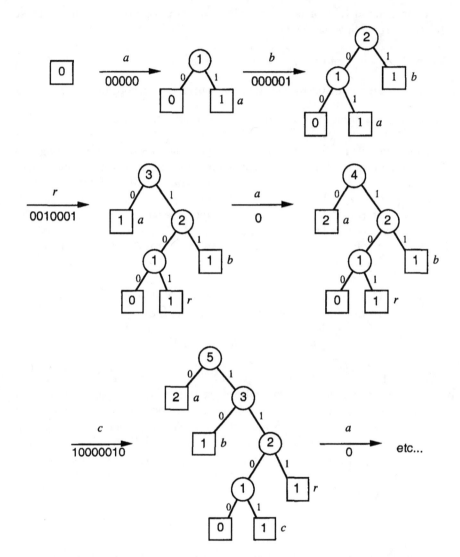

Figure 10.5: Dynamic Huffman compression of *abracadabra*.

text. This result comes from an increase of the alphabet on which the source *text* is written.

In the general case of factor encoding, a data compression scheme must solve the three following points:

- find the set F of factors that are to be encoded,

- design an algorithm to factorize the source text according to F,

- compute a code C in one-to-one correspondence with F.

When the text is given, the computation of such an optimal encoding is a NP-complete problem. The proof can be executed inside the following model of encoding. Let A be the alphabet of *text*. The encoding of *text* is a word in the form $d\#c$ where d ($\in A^*$) is supposed to represent the dictionary, $\#$ is a new letter (not in A), and c is the compressed text. The part c of the encoded string is written on the alphabet $A \cup \{1, 2, \ldots, n\} \times \{1, 2, \ldots, n\}$. A pair (i, j) occurring in c means a reference to the factor of length j that occurs at position i in d.

Example.   On $A = \{a, b, c\}$, let *text* $= aababbabbabbc$. Its encoding can be $hhahh\#aa(1, 4)a(0, 5)c$. The explicit dictionary is then

$$D = \{(babb, (1, 4)), (bbabb, (0, 5))\} \cup A \times A.$$

Within the above model, the length of $d\#c$ is the number of occurrences of both letters and pairs of integers that appear in it. For example, this length is 12 in the previous example. The search for a shortest word $d\#c$ that encodes a given *text* reduces to the SCS problem—the Shortest Common Superstring problem for a finite set of words—which is a classical NP-complete problem.

When the set F of factors is known, the main problem is to factorize the source s efficiently according to the elements of F, that is, to find factors $f_1, f_2, \ldots, f_k \in F$ such that

$$s = f_1 f_2 \ldots f_k.$$

The problem arises from the fact that F is not necessarily a unique decipherable code, several factorizations are often possible. It is important that the integer k be as small as possible, and a factorization is said to be an *optimal factorization* when k is minimal.

The simplest strategy for factorizing s is to use a greedy algorithm. The factorization is computed sequentially. Therefore, the first factor f_1 is naturally chosen as the longest prefix of s that belongs to F. And the decomposition of the remainder of the text is done iteratively in the same way.

Remark 1. If F is a set of letters or digrams ($F \subseteq A \cup A \times A$), the greedy algorithm computes an optimal factorization. The condition may seem quite restrictive, but in French, for example, the most frequent factors ("er" and "en") have length 2.

Remark 2. If F is a factor-closed set (all factors of words of F are in F), the greedy algorithm also computes an optimal factorization.

Another factorization strategy, called *semi-greedy* here, leads to optimal factorizations under broader conditions. Moreover, its time complexity is similar to the previous strategy.

Semi-greedy factorization strategy of s.

- let $m = \max\{|uv| : u, v \in F, \text{ and } uv \text{ is a prefix of } s\}$;

- let f_1 be an element of F such that $f_1 v$ is a prefix of s, and $|f_1 v| = m$, for some $v \in F$;

- set f_1 as the next factor;

- setting $s = f_1 s'$, iterate the same process on s'.

Example. Let $F = \{a, b, c, ab, ba, bb, bab, bba, babb, bbab\}$, and consider the greedy algorithm applied to $s = aababbabbabbc$. It produces the factorization

$$s = a.ab.ab.babb.ab.b.c$$

that contains 7 factors. The semi-greedy algorithm gives

$$s = a.a.ba.bbab.babb.c$$

that is an optimal factorization. Note that F is prefix-closed (prefixes of words of F are in F) after adding the empty word.

The interest in the semi-greedy factorization algorithm is due to the following lemma for which the proof is left as an exercise. As we shall see later in this section, the hypothesis of the set of factors F originates naturally for some compression algorithms.

Lemma 10.2 *If the set F is prefix-closed, the semi-greedy factorization strategy produces an optimal factorization.*

When the set F is finite, the semi-greedy algorithm may be realized with the help of a string-matching automaton (see Chapter 7). This leads to a linear-time algorithm to factorize a text.

Finally, if the set F is known, and if the factorization algorithm has been designed, the next step to consider when performing a whole compression process is to determine the code C. As for the statistical encodings of Sections 10.2 and 10.3, the choice of codewords associated with factors of F can take their frequencies into account. This choice can be made once for all, or in a dynamic way during the factorization of s. Here is an example of a possible strategy: the elements of F are put into a list and encoded by their position in the list. A move-to-front procedure realized during the encoding phase tends to attribute small positions, and thus short encodings, to frequent factors. The idea of encoding a factor by its position in a list is applied in the next compression method.

Factor encoding becomes even more powerful with an adaptive feature. It is realized by the Ziv-Lempel method (ZL method, for short). The dictionary is built during the compression process. The codewords of factors are their positions in the dictionary. Therefore, we regard the dictionary D as a mere sequence of words $(f_0, f_1, \dots)$. The algorithm encodes positions in the most efficient way according to the present state of the dictionary, using the function lg defined by:

$$\begin{cases} lg(1) = 1 & \text{and} \\ lg(n) = \lceil \log_2(n) \rceil & \text{for } n > 1. \end{cases}$$

Algorithm ZL; { Ziv-Lempel compression algorithm }
   { encodes source s on the binary alphabet }
     $D := \{\varepsilon\}$; $x := s\#$;
   while $x \neq \varepsilon$ **do begin**
       f_k = longest word of D such that $x = f_k ay$,
         for some $a \in A$;
       a := letter following f_k in x;
       write k on $lg|D|$ bits;
       write the initial codeword of a on $lg|A|$ bits;
       add $f_k a$ at the end of D;
      $x := y$;
   end

Example. Let $A = \{a, b, \#\}$. Assume that the initial codewords of letters are 00 for a, 01 for b, and 10 for $\#$. Let $s = aababbabbabb\#$. Then, the

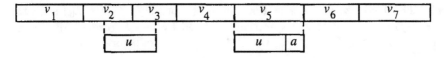

Figure 10.6: Efficient factorization of the source: v_5 is the shortest factor not occurring to its left.

decomposition of s is
$$s = a.ab.abb.abba.b.b\#$$
that leads to
$$c = 000\ 101\ 1001\ 1100\ 00001\ 10110.$$
After that, the dictionary D contains seven words:
$$D = (\varepsilon, a, ab, abb, abba, b, b\#).$$

Intuitively, ZL algorithm compresses the source text because it is able to discover some repeated factors within it. The second occurrence of such a factor is encoded only by a pointer onto its first position, which can save a large amount of space.

From the implementation point of view, the dictionary D in ZL algorithm can be stored as a word trie. The addition of a new word in the dictionary takes a constant amount of time and space.

There is a large number of possible variations on ZL algorithm. The study of this algorithm has been stimulated by its high performance in practical applications. Note, for example, that the dictionary built by ZL algorithm is prefix-closed, so the semi-greedy factorization strategy above may help to reduce the number of factors in the decomposition.

The model of encoding valid for that kind of compression method is a bit more general than the model introduced previously. In this model, the encoding c of the source text s is a word on the alphabet $A \cup \{1, 2, \ldots, n\} \times \{1, 2, \ldots, n\}$. A pair (i, j) occurring in c references a factor of s itself: i is the position of the factor, and j is its length.

Example. Again let $s = aababbabbabb\#$. It can be encoded by the word
$$c = a(0,1)b(1,2)b(3,3)ab(11,1)\#$$
of length 10, which corresponds to the factorization found by ZL algorithm.

The number of factors of the decomposition of the source text reduces if we consider a decomposition of the text similar to the f-factorization of Section

8.2. The factorization of s is a sequence of words $(f_1, f_2, \ldots, f_m)$ such that $s = f_1 f_2 \ldots f_m$ and that is iteratively defined by (see Figure 10.6): f is the shortest prefix of $f_i f_{i+1} \ldots f_m$ which does not occurs before in s.

Example. The factorization of $s = aababbabbabb\#$ is

$$(a, ab, abb, abbabb\#)$$

which accounts to encode s into

$$c = a(0,1)b(1,2)b(3,6)\#,$$

word of length 7 only, to be compared with the previous factorization.

Theorem 10.5 *The f-factorization of a string s can be computed in linear time.*

Proof. Use the directed acyclic word graph $DAWG(s)$, or the suffix tree $T(s)$. □

The number of factors appearing in the f-factorization of a text measures, in a certain sense, the complexity of the text. The complexity of sequences is related to the notion of entropy of a set of strings as stated below. Assume that the possible set of texts of length n (on the alphabet A) is of size $|A|^{hn}$ (for all length n). Then h (≤ 1) is called the entropy of the set of texts. For example, if the probability of appearance of a letter in a text does not depend on its position, then h is the entropy $H(A)$ considered in Section 10.2.

Theorem 10.6 *The number m of elements of the f-factorization of long enough texts is upper bounded by $hn/\log n$, for almost all texts.*

We end this section by reporting some experiments on compression algorithms. Table 10.1 gives the results. Rows are indexed by algorithms, and columns by types of text files. "Uniform" is a text on a 20-letter alphabet generated with a uniform and independent distribution of letters. "Repeated alphabet" is a repetition of the word $abc\ldots zABC\ldots Z$. Compression of the five files has been executed using the Huffman method, Ziv-Lempel algorithm (more precisely, COMPRESS command of UNIX), and the compression algorithm, called FACT, based on the f-factorization. Huffman method is the most efficient only for the file "Uniform", which is not surprising. For other files, the results obtained with COMPRESS and FACT are similar.

Sources	French text	C program	Uniform	Repeated alphabet
Initial length	62816	684497	70000	530000
Huffman	53.27 %	62.10 %	**55.58 %**	72.65 %
COMPRESS	**41.46 %**	34.16 %	63.60 %	2.13 %
FACT	47.43 %	**31.86 %**	73.74 %	**0.09 %**

Table 10.1: Sizes of some compressed files (best scores in bold).

Bibliographic notes

An extensive exposition of the theory of codes is given in [BP 85].

Elementary methods of data compression may be found in [He 87]. Practical programming aspects of data compression techniques are presented by Nelson in [Ne 95] and by Witten, Moffat and Bell in [WMB 99].

The construction of a minimal weighted tree is from Huffman [Hu 51]. The Shannon-Fano method is presented independently by Shannon (1948) and Fano (1949), see [BCW 90].

In the seventies, Faller [Fa 73] and Gallager [Ga 78] independently designed a dynamic data compression algorithm based on Huffman's method. Practical versions of dynamic Huffman coding have been designed by Cormack and Horspool [CH 84], and Knuth [Kn 85]. The precise analysis of sequential statistical data compression has been done by Vitter in [Vi 87], where an improved version is given.

NP-completeness of various questions on data compression can be found in the book of Storer [St 77]. The idea of the semi-greedy factorization strategy is from Hartman and Rodeh [HR 84]. And the dynamic factor encoding using the move-to-front strategy is by Bentley, Sleator, Tarjan, and Wei [BSTW 86]. This strategy is also used the method designed by Burrows and Wheeler (1999) and implemented by bzip.

In 1977, Ziv and Lempel designed the main algorithm of Section 10.4 (see [ZL 77] and also [ZL 88]). The notion of word complexity, and Theorem 10.6 appears in [LZ 76]. The corresponding linear-time computations are by Rodeh, Pratt, and Even [RPE 81] (with suffix trees) and Crochemore [Cr 83] (with suffix DAWG's). A large number of variants of Ziv-Lempel algorithm may be found in [BCW 90] and [St 88]. An efficient implementation of a variant of ZL algorithm is by Welch [We 84]. The experimental results of Section 10.4 are from Zipstein [Zi 92]. References and results relating compression ratios and entropy may be found in [HPS 92].

Considering words that do not occur in the source text instead of factors, Crochemore, Mignosi, Restivo, and Salemi (1998) have   designed the DCA method that has good performance [CMRS 01].

Some data compression methods do not use substitutions. A typical example is given by the application of arithmetic coding that often leads to higher efficiency because it can be combined with algorithms that evaluate or approximate the source probabilities. A software version of data compression based on arithmetic coding is by Witten, Neal, and Cleary [WNC 87]. It is not clear to whom application of arithmetic coding to compression should be attributed, see [BCW 90] for historical remarks on this point.

Chapter 11

Automata-theoretic approach

Finite automata can be considered both as simplified models of machines and as mechanisms used to specify languages. As machines, their only memory is composed of a finite set of states. In the present chapter, both aspects are considered and lead to different approaches to pattern matching. Formally, a (deterministic) automaton G is a sequence $(A, Q, init, \delta, T)$ where A is an input alphabet, Q is a finite set of states, $init$ is the initial state (an element of Q), and δ is the transition function. For reasons of economy we allow δ to be a partial function. The value $\delta(q, a)$ is the state reached from state q by the transition labelled by input symbol a, if any. The transition function extends in a natural way to all words, and, for the word x, $\delta(q, x)$ denotes, if it exists, the state reached after parsing the word x with the automaton from the state q. The set T is the set of accepting states, or terminal states of the automaton.

The automaton G accepts the language:

$$L(G) = \{x : \delta(init, x) \text{ is defined and belongs to } T\}.$$

The size of G, denoted by $size(G)$, is the number of transitions of G: number of pairs (q, a) (q is a state, a is a single symbol) for which $\delta(q, a)$ is defined. Another example for a useful size of G is the number of states denoted by $statesize(G)$.

Probably the most fundamental problem in this chapter is to construct in linear time a (deterministic) finite automaton G accepting the words ending by one pattern among a finite set of patterns, and gives a representation of G of linear size, independently on the size of the alphabet. The use of the automaton leads to a linear-time searching algorithm.

163

The background of the previous question is the regular-expression matching, in which the finite set is extended into a regular expression e. It is still possible to build a linear-size automaton equivalent to the expression, but it is no longer deterministic. This yields a slower algorithm to localize patterns as stated in the next theorem.

Theorem 11.1 *The recognition of patterns specified by a regular expression e in a text text can be realized in time $O(size(e) \times |text|)$ and space $O(size(e))$.*

If the automaton is made deterministic, its size may grow exponentially but the searching time no longer depends on the expression as stated next.

Theorem 11.2 *The recognition of patterns specified by a regular expression e in a text text can be realized in time $O((|A| \times 2^{size(e)}) + |text|)$ and space $O(|A| \times 2^{size(e)})$, where A is the set of letters occurring effectively in e.*

The proof of the two above theorem may be found in books on automata and compiling (see bibliographic references).

11.1　Aho-Corasick automaton

We denote by $SMA(pat)$, for *String-Matching Automaton*, a (deterministic) finite automaton G accepting the set of all words containing pat as a suffix. In similar way, we denote by $SMA(\Pi)$ an automaton accepting the set of all words having a word of the finite set of words Π as a suffix. In other words, noting A^* the set of all words on the alphabet A,

$$L(SMA(pat)) = A^*pat \text{ and } L(SMA(\Pi)) = A^*\Pi.$$

In this section we present a construction of the minimal finite automaton $G = SMA(pat)$, where minimality is understood based on the number of states of automata.

Unfortunately, the total size of G depends heavily upon the size of the alphabet. We show how to construct these automata in linear time (with respect to the output).

With these automata, the following real-time algorithm SMA can be applied to solve the string-matching problem (for one or many patterns). The algorithm outputs a string of 0's and 1's that locates all occurrences of the pattern in the text (1's mark the end position of occurrences of the pattern). The algorithm does not use the same model of computation as the algorithms of Chapters 3 and 4 do. There, the elementary operation used by algorithms

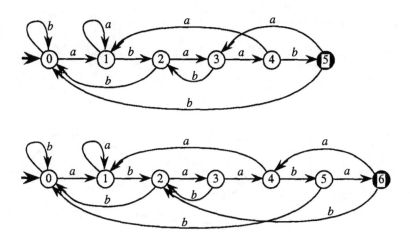

Figure 11.1: One step in the construction of an SMA—from $SMA(abaab)$ to $SMA(abaaba)$—unfolding the a-transition from state 5. Terminal states are black.

(MP, BM, and their variations) is letter comparison. Here, the basic operation is branching (computation of a transition).

```
Algorithm SMA { real-time transducer for string-matching }
    state := init; read(symbol);
    while symbol ≠ end-marker do begin
        state := δ(state, symbol);
        if state in T then
            write(1); { it reports an occurrence of the pattern }
        else write(0);
        read(symbol);
    end
```

We start with the case of only one pattern pat. We show how to build the minimal automaton $SMA(pat)$. The function $buildSMA$ builds $SMA(pat)$ sequentially. The core of the construction consists, for each letter a of the pattern, in unfolding the a-transition from the last created state t. This is illustrated on pattern $ababa$ in Figure 11.1.

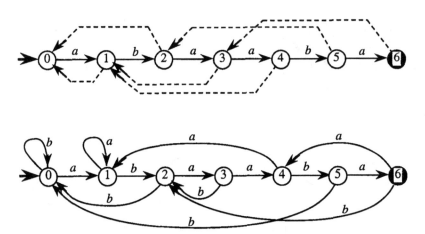

Figure 11.2: The function *Bord* (dotted arrows) of pattern *abaaba*, and the automaton *SMA(abaaba)*.

```
function buildSMA(pat): automaton;
    create a state init; terminal := init;
    for all b in A do δ(init, b) := init;
    for a := first to last letter of pat do begin
        temp := δ(terminal, a); δ(terminal, a) := new state x;
        for all b in A do δ(x, b) := δ(temp, b);
        terminal := x;
    end;
    return (A, set of created states, δ, init, {terminal});
```

Lemma 11.1 *Algorithm buildSMA constructs the automaton SMA(pat) in time $O(m.|A|)$. The (minimal) automaton SMA(pat) has $(m + 1)|A|$ transitions, and $m + 1$ states.*

Proof. The proof is left as an exercise.   □

There is an alternative construction of the automaton *SMA(pat)* that shows the relation between SMA's and the MP-like algorithm of Chapter 3. Once we have computed the failure table *Bord* for pattern *pat*, the automaton *SMA(pat)* can be constructed as follows. We first define $Q = \{0, 1, \ldots, m\}$, $T = \{m\}$, $init = 0$. The transition function (table) δ is computed by the algorithm below. Figure 11.2 simultaneously displays the failure links (arrows going to the left) and $SMA(abaaba)$.

In some sense, we can consider that table *Bord* represents the transition function δ of the automaton $SMA(pat)$. Then, *MP* algorithm becomes a mere simulation of algorithm *SMA* above. In the simulation, branching operations are substituted by letter comparisons. This remark is indeed the basis of Simon's algorithm. The representation of $SMA(pat)$ by a failure function makes the size of the representation independent of the alphabet without increasing the total time complexity of the search phase.

Algorithm { computes the transition function of $SMA(pat)$ }
            { assuming that table *Bord* is already computed }
    for all a in A **do** $\delta(0, a) := 0$;
    if $m > 0$ **then** $\delta(0, pat[1]) := 1$;
    for $i := 1$ **to** m **do**
        for all a in A **do**
            if $i < m$ **and** $a = pat[i + 1]$ **then** $\delta(i, a) := i + 1$
            else $\delta(i, a) := \delta(Bord[i], a)$;

The algorithm above shows that the transition function of $SMA(pat)$ can be computed from the failure table *Bord*.

We next apply the same strategy to the recognition of a finite set of patterns. Assume that we have a set Π of r patterns. The i-th pattern is denoted by p_i. Let m be the sum of lengths of all patterns. We no longer try to build the minimal string-matching automaton corresponding to the problem. Therefore, $SMA(\Pi)$ is not necessarily the minimal (deterministic) automaton of the language $A^*\Pi$, as it is when Π contains only one pattern.

To construct $SMA(\Pi)$, we first consider a tree (a word trie) $Tree(\Pi)$ in which the branches are labelled by elements of Π. The nodes of $Tree(\Pi)$ are identified with prefixes of words in Π. The root is the empty word ε. The father of a non-empty prefix xa (a a letter) is the prefix x. We write $father(xa) = x$, and $child(x, a) = xa$. Nodes of $Tree(\Pi)$ are considered as states of an automaton, and they are marked terminal or non-terminal. A node is marked terminal if the word it represents is in the set Π. All leaves are terminal states, but it may also occur that some internal node is also a terminal state. This happens when a pattern is a proper prefix of another pattern. Figure 11.3 displays $Tree(\{ab, babb, bb\})$. When applied to a set of string Π, the algorithm *buildSMA* below builds an automaton $SMA(\Pi)$ from the tree $Tree(\Pi)$. The states of the automaton are the nodes of the tree. The algorithm essentially transforms and completes the relation *child* of $Tree(\Pi)$ into the transition δ of $SMA(\Pi)$. The algorithm is very similar to the case of a

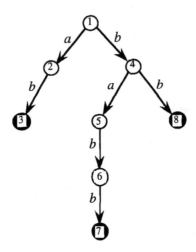

Figure 11.3: The trie of set $\{ab, babb, bb\}$. Terminal nodes are black (3, 7, 8).

single word. Here, no state is created because states are taken from the input tree.

There is, however, a delicate point involved in the computation of terminal states. It may occur that some word of Π is an internal factor of another word of the set. This is the case for the set $\{ab, babb, bb\}$ considered in Figure 11.3 because ab is an internal factor of $babb$. Node 6 of $Tree(\{ab, babb, bb\})$ becomes a terminal state in $SMA(\{ab, babb, bb\})$, because bab ends with the word ab that is in the set (see Figure 11.4). More generally, this happens during the construction when the clone of node x, namely node $temp$ in the algorithm, is itself a terminal node.

Lemma 11.2 *The algorithm buildSMA applied to a set Π of strings builds a deterministic automaton $SMA(\Pi)$ having the same set of nodes as the trie $Tree(\Pi)$. It runs in time $O(statesize(Tree(\Pi)) \times |A|)$.*

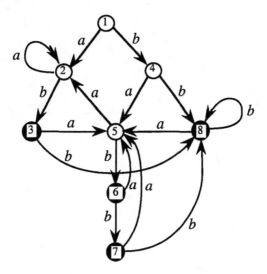

Figure 11.4: The automaton $SMA(\Pi)$ for $\Pi = \{ab, babb, bb\}$. Note that node 6 is terminal.

---

function $buildSMA(\Pi)$: automaton;
{ *father* and *child* links refer to the tree $Tree(\Pi)$ }
{ states of $SMA(\Pi)$ are the nodes of $Tree(\Pi)$ }
    if *root* is a terminal node **then** $T := \{root\}$ **else** $T := \emptyset$;
    for all b in A **do** $\delta(root, b) := root$;
    for all non-root nodes x of $Tree(\Pi)$ in bfs order **do begin**
        $t := father(x)$; $a :=$ the letter such that $x = child(t, a)$;
        $temp := \delta(t, a)$; $\delta(t, a) := x$;
        if x **or** $temp$ are terminal nodes **then** add x to T;
        for all b in A **do**
            if $child(temp, b)$ is defined **then** $\delta(x, b) := child(temp, b)$
            else $\delta(x, b) := \delta(temp, b)$;
    end;
    return$(A$, nodes of $Tree(\Pi)$, δ, *root*, $T)$;

The automaton $SMA(\Pi)$ can also be built from scratch without a previous computation of $Tree(\Pi)$. Branches are unfolded as in the case of one pattern. Patterns are processed simultaneously, all prefixes of the same length at a time, which correspond to the breath-first-search order applied on $Tree(\Pi)$ by the algorithm *buildSMA*. Figure 11.5 illustrates this alternative strategy for

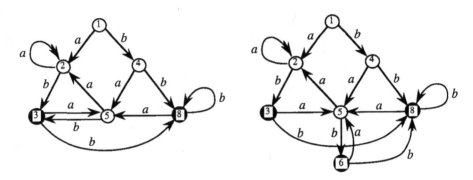

Figure 11.5: One step of another possible algorithm for the construction of $SMA(\{ab, babb, bb\})$. Node 6 is a clone of $\delta(5, b) = 3$ (on the left). Transitions on node 6 are the same as those on node 3.

building $SMA(\Pi)$.

   With the automaton $SMA(\Pi)$ built by the previous algorithm, searching a text for occurrences of patterns that are in Π can be realized by the algorithm SMA. The search is then performed in real time, and the space required to store the automaton is $O(statesize(Tree(\Pi)) \times |A|)$. Again, it is possible to represent the automaton $SMA(\Pi)$ with a failure function. The advantage in doing so is to represent the automaton within space $O(statesize(Tree(\Pi)))$, quantity that is independent of the alphabet. The search then becomes analogous to MP algorithm of Chapter 3.

   Continuing the analogy, a function $Bord$ related to Π can then be defined as follows. For a non-empty word u,

   $Bord(u) =$ longest proper suffix of u that is a prefix of some pattern in Π.

We also denote by $Bord$ the failure table defined on nodes of $Tree(\Pi)$ (except on the root) (see Figure 11.6). The relation used by the next algorithm that computes table $Bord$ is

$$\begin{cases} Bord[ta] = Bord^k[t]a & \text{for the smallest } k \text{ such that } Bord^k[t]a \in Trie(\Pi), \\ Bord[ta] = \varepsilon & \text{otherwise,} \end{cases}$$

where t is a node of $Tree(\Pi)$ different from the root.

   Note that the algorithm also marks nodes as terminal in the same situation as that explained for the direct construction of $SMA(\Pi)$.

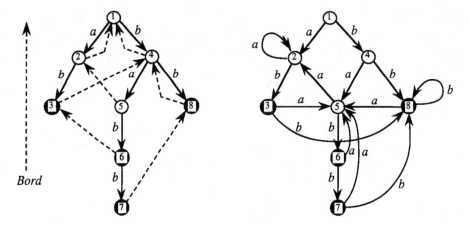

Figure 11.6: The tree $Tree(\Pi)$ with suffix links $Bord$ (left), and the automaton $SMA(\Pi)$ (right), for $\Pi = \{ab, babb, bb\}$.

---

procedure *compute-Bord*; { failure table on $Tree(\Pi)$ }
{ *father* and *child* refer to the tree $Tree(\Pi)$ }
{ *Bord* is defined on nodes of $Tree(\Pi)$ }
    set $Bord[\varepsilon]$ as undefined;
    for all a in $Tree(\Pi)$ **do** $Bord[a] := \varepsilon$;
    for all nodes x of $Tree(\Pi)$, $|x| > 2$, in bfs order **do begin**
        $t := father(x)$; $a :=$ the letter such that $x = child(t, a)$;
        $z := Bord[t]$;
        while z is defined **and** za is not in $Tree(\Pi)$ **do** $z := Bord[z]$;
        if za is in $Tree(\Pi)$ **then** $Bord[x] := za$ **else** $Bord[x] := \varepsilon$;
        if $Bord[u]$ is terminal **then** mark x as terminal;
    end

The time complexity of the above algorithm is proportional to m, the total size of Π. The analysis is similar to that of computing $Bord$ for a single pattern. It is sufficient to estimate the total number of all executed statements $z := Bord[z]$. This statement can again be treated as deleting some items from a store and z as the number of items. Let us fix a path π of length k from the root to a leaf. Using the store principle it is easy to prove that the total number of insertions (increases of z) into the store for nodes of π is bounded by k, hence, the total number of deletions (executing $z := Bord[z]$) is also bounded by k. If we sum this over all paths we then get the total length m of all patterns in Π.

We can again base the construction of an automaton $SMA(\Pi)$ on the failure table of $Tree(\Pi)$. The transition function is defined on nodes of the tree as shown by the following algorithm.

Algorithm { computes transition function for $SMA(\Pi)$ }
    for all a not in $Tree(\Pi)$ **do** $\delta(\varepsilon, a) := \varepsilon$;
    for all nodes x of $Tree(\Pi)$, $|u| > 0$, in bfs order **do**
        for all a in A **do**
            if xa is in $Tree(\Pi)$ **then** $\delta(x, a) := xa$
            else $\delta(x, a) := \delta(Bord[x], a)$;

---

function $AC(Tree(\Pi); text)$: boolean;
  { Aho-Corasick multi-pattern matching }
  { uses the table $Bord$ on $Tree(\Pi)$ }
    $state := root$; read($symbol$);
    while $symbol \neq$ end-marker **do begin**
        while $state$ is defined **and** $child(state, symbol)$ is undefined **do**
            $state := Bord[state]$;
        if $state$ is undefined **then** $state := root$
        else $state := child(state, symbol)$;
        if $state$ is terminal **then return** true;
        read($symbol$);
    end;
    return false;

Algorithm AC is the version of algorithm MP for several patterns. It is a straightforward application of the notion of the failure table (namely $Bord$). The preprocessing phase of AC algorithm is the procedure *compute-Bord*.

Terminal nodes of $SMA(\Pi)$ can be assigned numbers corresponding to patterns in Π. Hence, the automaton can produce in real-time numbers that correspond to those patterns ending at the last scanned position of the text. If no pattern occurs, 0 is written. This proves the following statement.

Theorem 11.3 *The string-matching problem for a finite number of patterns can be solved in time $O(n + m)$. The additional memory is of size $O(m)$.*

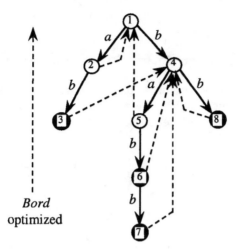

Figure 11.7: Optimized table *Bord* on *Tree*({*ab*, *babb*, *bb*}). Note that the table is undefined on node 4.

The table *Bord* used in *AC* algorithm can be improved in the following manner. Assume, for example, that during the search, node 6 of Figure 11.6 has been reached, and that the next letter is *a*. Since node 6 has no *a*-son in the tree, *AC* algorithm iterates function *Bord* from node 6. It successively finds nodes 3 and node 4 where the iteration stops. It is clear that the test on node 3 is useless in any situation because this node has no son at all. On the contrary, node 4 plays its role because it has an *a*-son. Some iterations of the function *Bord* can be precomputed on the tree. Figure 11.7 shows the result of the transformation. The optimization does not change the worst-case behavior of *AC* algorithm.

11.2 Determinizing automata

We consider a certain set S of patterns represented in a succinct way by a deterministic automaton G with n states. The set S is the language $L(G)$ accepted by G. Typical examples of sets of patterns are $S = \{pat\}$, a singleton, and $S = \{pat_1, pat_2, \ldots, pat_r\}$, a finite set of words. In the first example, the structure of G is mainly done by the line of consecutive positions in *pat*. The rightmost state (position) is the only terminal state. In the second example, the structure of G is the tree of prefixes of patterns. All leaves of the tree are terminal states, and some internal nodes can also be terminal if a pattern is a

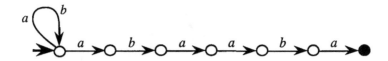

Figure 11.8: The non-deterministic pattern-matching automaton for *abaaba*. The powerset construction gives the automaton *SMA(abaaba)* of Figure 11.2. An efficient case of the powerset construction.

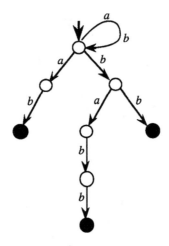

Figure 11.9:   The non-deterministic pattern-matching automaton for the set {*ab, babb, bb*}.   The powerset construction gives the automaton *SMA*({*ab, babb, bb*}) of Figure 11.4. An efficient case of the powerset construction.

prefix of another pattern of the set.

We can transform G into a non-deterministic pattern-matching automaton, denoted by *loop(G)*, that accepts the language of all words having a suffix in $L(G)$. The automaton *loop(G)* is obtained by adding a loop on the initial state, for each letter of the alphabet. The automata *loop(G)* for the two examples of cases mentioned above are presented in Figure 11.8 and Figure 11.9 respectively. The actual non-determinism of the automata appears only on the initial state.

We can apply the classical powerset construction (see [HU 79], for example) to a non-deterministic automaton *loop(G)* to get an equivalent deterministic automaton. It appears that, in the two cases of one pattern and of a finite

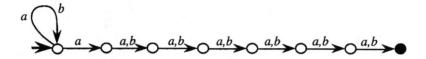

Figure 11.10: A non-efficient case of the powerset construction. The displayed automaton results by adding loops onto the initial state of the deterministic automaton accepting words of length 7 that start with letter a. The smallest equivalent deterministic automaton has 128 states.

set of patterns, by doing so we obtain efficiently deterministic string-matching automata. In the powerset construction, only those subsets that are accessible from the initial subset are considered.

However, the idea of using $loop(G)$ and the powerset construction on it altogether is not always efficient. In Figure 11.10 a non-efficient case is presented. It is not hard to become convinced that the deterministic version of $loop(G)$, which has a tree-like structure, cannot have less than 2^7 states. Extending the example shows that there exists a non-deterministic automaton, even of the form $loop(G)$, with $n + 1$ states that is transformed into an equivalent deterministic automaton having 2^n states. And that is has no smaller equivalent deterministic automaton.

Similarly we can transform the deterministic automaton G accepting one word pat into a non-deterministic automaton $FAC(G)$ accepting the set $Fac(pat)$ of factors of pat (see Figure 11.11). The transformation can be done simply by making all states simultaneously initial and terminal states. But we prefer modifying it by creating additional edges from the single initial state to all other states, edges that are labelled by all letters. It happens again that we find an efficient case of the powerset construction if we start with the deterministic automaton accepting just a single word. The powerset construction gives the smallest deterministic automaton accepting the suffixes of pat.

The powerset construction applied on automata of the form $FAC(G)$ is not always efficient (see Figure 11.12). If we take as G the deterministic automaton with $2n+3$ states accepting the set $S = (a+b)^n a(a+b)^n c$, then the automaton $FAC(G)$ also has $2n + 3$ states. But the smallest deterministic automaton accepting the set of all factors of words in S has an exponential number of states.

Combining $loop$ and FAC sometimes yields an efficient powerset construction. This is done implicitly in Section 6.2 or when a DAWG is used as a pattern-matching machine. There, the failure function defined on the automaton $DAWG(pat)$ serves to represent the automaton $loop(DAWG(pat))$. The

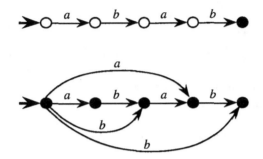

Figure 11.11: An efficient case of the powerset construction. Applied to $FAC(G)$ accepting $Fac(text)$, it gives the smallest automaton accepting the suffixes of pat; it has a linear number of states.

overall result is efficient because both steps—from pat to $DAWG(pat)$ and from $DAWG(pat)$ to $loop(DAWG(pat))$—are. However, in general, the whole determinization process is inefficient.

11.3   Two-way pushdown automata

The two-way deterministic pushdown automaton (2dpda) is an abstract model of linear-time computations on texts. A 2dpda G is essentially a deterministic pushdown finite-state machine (see [HU 79]) that differs from the standard model in its ability to move the input head in two directions.

The possible actions of the automaton are: changing the current state of the finite control, moving the head by one position, and locally changing the contents of the stack "near" its top. For simplicity, we assume that each change of the contents of the stack is of one of two types:

- push(a)—pushing a symbol a onto the stack;

- pop—popping one symbol off the stack.

The automaton has access to the top symbol of the stack and to the symbol in front of the two-way head. We also assume that there are special left and right end-markers on both ends of the input word. The output of such an abstract algorithm is "true" iff the automaton stops in an accepting state. Otherwise, the output is "false." Initially the stack contains a special "bottom" symbol, and we assume that at the final moment of acceptance the stack also contains

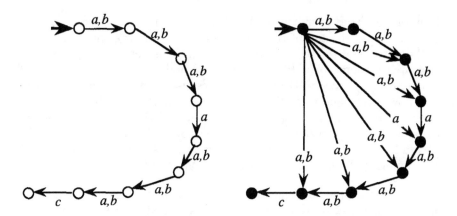

Figure 11.12: A non-efficient case of the powerset construction. The automaton G (on the left) accepts $(a+b)^n a(a+b)^n c$, for $n = 3$. A deterministic version of $FAC(G)$ (on the right) has, in this case, an exponential number of states.

one element. When the stack is empty the automaton stops (the next move is undefined).

A problem solved by a 2dpda can be treated abstractly as a formal language L consisting of all words for which the answer of the automaton is "true" (G stops in an accepting state). We say also that G accepts the language L. The string-matching problem, for a fixed input alphabet, can be also interpreted as the formal language:

$$L_{sm} = \{pat\&text : pat \text{ is a factor of } text\}.$$

This language is accepted by some 2dpda and this property gives an automata-theoretic approach to linear-time string-matching, because, as we shall see later, a 2dpda can be simulated in linear time. Historically, in fact, it was one of the first methods used for the (theoretical) design of a linear-time string-matching algorithm.

Lemma 11.3 *There is a 2dpda accepting the language L_{sm}.*

Proof. We define a 2dpda G for L_{sm}. It simulates the naive string-matching algorithm *brute-force1* (see Chapter 3). At a given stage of the algorithm, we start at a position i in the text *text*, and at the position $j = 1$ in the pattern *pat*. The pair (i, j) is replaced by the pair (stack, j), where j is the position of the input head in the pattern (and has exactly the same role as j in algorithm *brute-force1*). The contents of the stack is $text[i..n]$ with $text[i+1]$ at the top.

The automaton tries to match the pattern starting from position $i + 1$ in the text. It checks if the top of the stack equals the current symbol at position j in the pattern; if so, then $(j = j + 1, pop)$. This action is repeated until a mismatch is found, or until the input head points on the marker "&." In the latter case, G accepts. Otherwise, it goes on to the next stage. The stack should now correspond to $text[i + 1..n]$ and $j = 1$. It is not difficult to reach such a configuration. The stack is reconstructed by shifting the input head back while simultaneously pushing scanned symbols of pat (that have been matched successfully against the pattern). This shows that the algorithm $brute\text{-}force1$ can be simulated by a 2dpda, and completes the proof.   □

Similarly to the previous simulation, the problem of finding the prefix palindromes of a string, and several other problems related to palindromes, can be interpreted as formal languages. Here is a sample:

$$L_{prefpal} = \{ww^{R}u : u, w \in A^{*}, w \text{ non-empty word}\},$$

$$L_{prefpal3} = \{ww^{R}uu^{R}vv^{R}z : w, v \in A^{*}, w, u, v \text{ non-empty words}\},$$

$$L_{pal2} = \{ww^{R}uu^{R} : w, u, v \in A^{*}, w, u \text{ non-empty words}\}.$$

All these languages related to symmetries are accepted by 2dpda. We leave it as an exercise to construct appropriate 2dpdas.

The main feature of 2dpdas is that they correspond to some simple linear-time algorithms. Assume that we are given a 2dpda G and an input word w of length n. The size of the problems is n (the static description of G has constant size). Then, it is proven below that testing whether w is in $L(G)$ takes linear time. We present the concepts that lead to the proof of the result.

The key concept for 2dpda is the consideration of *top configurations* also called *surface configurations*. The top configuration of a 2dpda retains its top element from the stack only. The first basic property of top configurations is that they contain sufficient information for the 2dpda to choose the next move. The entire configuration consists of the current state, the present contents of the stack, and the position of the two-way input head. Unfortunately, there are too many such configurations (potentially infinite, as the automaton can loop while making push operations). It is easy to see that in every accepting computation the height of the stack is linearly bounded. But this does not help very much because there is an exponential number of possible contents of the stack of linear height. The second basic property of top configurations is that their number is linear in the size of the problem n.

Formally, a top configuration is a tuple $C = (state, top, symbol, position)$. The linearity of the set of top configurations obviously follows from the fact that the number of states and the number of top symbols (elements of the stack alphabet) are bounded by a constant: they do not depend on n.

We can classify top configurations according to the type of next move of the 2dpda as *pop configurations* and *push configurations*. For a top configuration C, we define $Term[C]$ as a pop configuration C' accessible from C after some number (possibly zero) of moves that do not pop the top symbol C. If there is no such C' then $Term[C]$ is undefined. Assume w. l. o. g. that the accepting configuration is a pop configuration and that the stack is a simple element.

The theorem below is the main result related to 2dpdas. It is surprising in view of the fact that the number of moves is usually bigger than that of top configurations. In fact, a 2dpda can make an exponential number of moves and halt. But the result simply shows that shortcuts are possible.

Theorem 11.4 *If the language L is accepted by a 2dpda, then there is a linear-time algorithm to test whether x belongs to L (for fixed-size alphabets).*

Proof. Let G be a 2dpda, and let w be a given input word of length n. Let us introduce two functions acting on top configurations:

- $P1(C) = C'$, where C' results from C by a push move; this is defined only for push configurations.

- $P2(CI, C2) = C'$, where C' results from $C2$ by a pop move, and the top symbol of C' is the same as in $C1$ ($C2$ determines only the state and the position).

Let POP be the boolean function defined by: $POP(C) =$ true iff C is a pop configuration. All these functions can be evaluated in constant time by a random access machine using the (constant-sized) description of the automaton.

It is sufficient to compute in linear time the value of $Term[C0]$ (or find that it is undefined), where $C0$ is an initial top configuration. According to our assumptions simplifying 2dpdas, if G accepts, then $Term[C0]$ is defined. Assume that initially all entries of the table $Term$ contain a special value "not computed."

We start with the assumption that G never loops and ends with a one-element stack. In fact, if the move of G is at some moment undefined we can assume that it goes to a state in which all the symbols in the stack are successively popped.

Algorithm { linear-time simulation of the halting 2dpda }
    for all configuration C **do** *onstack*$[C]$:=false;
    return $Comp(C0)$;

```
function Comp(C);
    if Term[C] = "not computed" then
        if POP(C) then Term[C] := C
        else Term[C] := Comp(P2(C, Comp(P1(C))));
    return Term[C];
```

The correctness and time linearity of the algorithm above are obvious in the case of a halting automaton. The statement *"Term[C] := ..."* is executed at most once for each C. Hence, only a linear number of push moves (using function $P1$) are applied.

```
Algorithm Simulate1;
    { a version of the previous algorithm that detects loops }
        for all configuration C do onstack[C] :=false;
        return Comp1(C0);
```

```
function Comp1(C);
    { returns Term[C] if defined, 'false' otherwise, C1 is a local variable }
    if Term[C] = "not computed" then begin
        C1 := P1(C);
        if onstack[C1] then return false { loop }
        else onstack[C1] := true;
        C1 := Comp(C1); onstack[C1] := false; { pop move };
        C1 := P2(C, C1);
        if onstack[C1] then return false { loop } else begin
            onstack[C] :=false; onstack[C1] :=true
        end;
        Term[C] := Comp(C1)
    end;
    return Term[C];
```

The algorithm *Simulate1* is for the general case: it also detects a possible looping of G for a given input. We use the table *onstack* initially consisting of "false" values. Whenever we make a push move we then set *onstack[C1]* to true for the current top configuration $C1$, and whenever a pop move is made we set *onstack[C1]* to false. The looping is detected if the automaton tries to put a configuration that is already on the (virtual) stack of top configurations. And if there is a loop, such a situation occurs.

The algorithm *Simulate1* has also a linear-time complexity by the same argument as in the case of halting 2dpdas. This completes the proof. $\square$

Sometimes it is quite difficult to design a 2dpda for a given language L, even if we Know that such 2dpda exists. An example of such a language L is:

$$L = \{1^n : n \text{ is the square of an integer}\}.$$

A 2dpda for this language can be constructed by a method presented in [Mo 85]. It is much easier to construct 2dpdas for the following languages:

$$\{a^n b^m : m = 2^n\}, \quad \{a^n b^m : m = n^4\}, \quad \{a^n b^m : m = \log^*(n)\}.$$

But it is not known whether there is a 2dpda accepting the set of even palstars, or the set of all palstars. In general, there is no good technique for proving that a specific language is not accepted by any 2dpda. In fact the "P= NP?" problem can be reduced to the question: does a 2dpda exist for a specific language L? There are several examples of such languages. Generally, 2dpdas are used to give alternative formulations of many important problems in complexity theory.

Bibliographic notes

Two "simple" pattern-matching machines are discussed by Aho, Hopcroft, and Ullman in [AHU 74] and, in the case of many patterns, by Aho and Corasick in [AC 75]. The algorithms first compute failure functions, and then the automata. The constructions given in Section 11.1 are direct constructions of the pattern-matching machines. The algorithm of Aho and Corasick is implemented by the command "fgrep" of UNIX system. A version of *BM* algorithm adapted to the search for a finite set of patterns was first sketched by Commentz-Walter [Co 79]. The algorithm was completed by Aho [Ah 90]. Another version of Commentz-Walter's algorithm is presented in [BR 90]. An algorithm for multiple string searches is presented in Crochemore *et al.* [C-R 93], where experiments on the real behavior of the algorithm are presented.

The determinization of automata can be found in the standard textbook of Hopcroft and Ullman [HU 79]. The question of efficient determinization of automata is from Perrin [Pe 90]. This paper is a survey on the main properties and discoveries about automata theory.

Chapter 12

Approximate pattern matching

In practical pattern-matching applications, the exact matching is not always pertinent. It is often more important to find objects that match a given pattern in a reasonably approximate way. In this chapter, approximation is measured mainly by the so-called edit distance: the minimal number of local edit operations needed to transform one object into another. The analogue for DNA sequences is called the *alignment problem* (see Figure 12.1). Algorithms are mainly based on the algorithmic method called *dynamic programming*. We also present problems strongly related to the notion of edit distance, namely, the computation of longest common subsequences, string matching allowing errors, and string matching with don't care symbols.

12.1 Edit distance

An immediate question arising in applications is how to test the equality of two strings allowing some errors. The errors correspond to differences between

Figure 12.1: Alignment of two DNA sequences showing the operations of changes, insertions ("–" in top line), and deletions ("–" in bottom line).

the two words. We consider in this section three types of differences between
two strings x and y:

change: symbols at corresponding positions are distinct,

insertion: a symbol of y is missing in x at a corresponding position,

deletion: a symbol of x is missing in y at a corresponding position.

We require the minimum number of differences between x and y. We translate
this as the smallest possible number of operations (change, deletion, insertion)
to transform x into y. This is called the *edit distance* between x and y, and
denoted by $edit(x, y)$. It is clear that it is a distance between words, in the
mathematical sense. This means that the following properties are satisfied:

- $edit(x, y) \geq 0$,

- $edit(x, y) = 0$ iff $x = y$,

- $edit(x, y) = edit(y, x)$ (symmetry),

- $edit(x, y) \leq edit(x, z) + edit(z, y)$ (triangle inequality).

The symmetry of *edit* comes from the duality between deletions and insertions:
a deletion of the letter a of x in order to get y corresponds to an insertion of
a into y to get x.

Example. The text $x = wojtk$ can be transformed into $y = wjeek$ using one
deletion, one change and one insertion. This shows that $edit(wojtk, wjeek) \leq$
3, because it uses three operations. In fact, this is the minimum number of
edit operations to transform $wojtk$ into $wjeek$.

From now on, we consider that words x and y are fixed. The length of x is
m, and the length of y is n, and we assume that $n \geq m$. We define the table
$EDIT$ by:

$$EDIT[i, j] = edit(x[1..i], y[1..j])$$

for $0 < i \leq m$ and $0 < j \leq n$. The boundary values are defined as follows (for
$0 \leq i \leq m, 0 \leq j \leq n$):

$$EDIT[0, j] = j, \quad EDIT[i, 0] = i.$$

There is a simple formula for computing other elements.

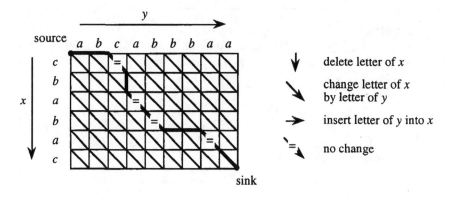

Figure 12.2: The path corresponds to the sequence of edit operations: *insert(a)*, *insert(b)*, *delete(b)*, *insert(b)*, *insert(b)*, *change(c, a)*.

(*) *Dynamic Programming Recurrence:*

$$EDIT[i, j] = \min(EDIT[i - 1, j] + 1, EDIT[i, j - 1] + 1,$$
$$EDIT[i - 1, j - 1] + \partial(x[i], y[j]),$$

where $\partial(a, b) = 0$ if $a = b$, and $\partial(a, b) = 1$ otherwise.

The formula reflects the three operations, deletion, insertion, and change, in that order.

There is a graph theoretical formulation of the editing problem. We consider the *grid graph* also called the *alignment graph*, denoted by G. It is composed of nodes (i, j) (for $0 \le i \le m$, $0 \le j \le n$).

The node $(i - 1, j - 1)$ is connected to the three nodes $(i - 1, j), (i, j - 1), (i, j)$ when they are defined (i.e., when $i \le m$, or $j \le n$).

Each edge of the grid graph has a weight corresponding to the recurrence (*). The edges from $(i - 1, j - 1)$ to $(i - 1, j)$ and $(i, j - 1)$ have weight 1, as they correspond to the insertion and deletion of a single symbol, respectively. The edge from $(i - 1, j - 1)$ to (i, j) has weight $\partial(x[i], y[j])$.

Figure 12.2 shows an example of grid graph for words *cbabac* and *abcabbbaa*.

The edit distance between words x and y equals the length of a least weighted path in this graph from the source $(0, 0)$, left upper corner, to the sink (m, n), right bottom corner.

```
function edit(x, y) { computation of edit distance }
{ |x| = m, |y| = n, EDIT is a matrix of integers }
   for i := 0 to m do EDIT[i, 0] := i;
   for j := 1 to n do EDIT[0, j] := j;
   for i := 1 to m do
      for j := 1 to n do
         EDIT[i, j] = min(EDIT[i − 1, j] + 1, EDIT[i, j − 1] + 1,
                          EDIT[i − 1, j − 1] + ∂(x[i], y[j]));
   return EDIT[m, n];
```

The algorithm above computes the edit distance of strings x and y. It stores and computes all values of the matrix $EDIT$, although only one entry, $EDIT[m, n]$, is required. This serves to save time, and this feature is called the *dynamic programming method*. Another possible algorithm to compute $edit(x, y)$ could be to use the classical Dijkstra's algorithm for shortest paths in the grid graph.

Theorem 12.1 *Edit distance of two words of lengths m and n can be computed in $O(mn)$ time using $O(\min\{m, n\})$ additional memory.*

Proof. The time complexity of the algorithm above is obvious. The space complexity results from the fact that we do not have to store the entire table. The current and the previous columns (or lines) are sufficient for carrying out computations. □

We can assign specific costs to edit operations, depending on the type of operations and on the type of symbols involved. Such a generalized edit distance can be computed using a formula analogous to equation (*).

As noted in the proof of the previous theorem, the whole matrix $EDIT$ does not need to be stored (only two columns are sufficient at a given step) in order to compute only $edit(x, y)$. However, we can keep it in memory if we want to compute a shortest sequence of edit operations transforming x into y. This is essentially done by tracing back in the matrix how each value has been obtained.

12.2   Longest common subsequence problem

In this section, we consider a problem that illustrates a particular case of the edit distance problem of the previous section. This is the example of computing

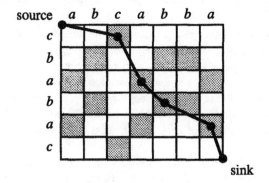

Figure 12.3: The size of the longest subsequences is the maximal number of shaded boxes (associated with matches) on a monotonically decreasing path from source to sink. Compare with Figure 12.4.

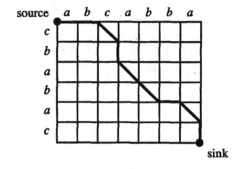

Figure 12.4: Assigning cost 2 diagonal edges, the length of the path is $m + n$. Diagonal edges correspond to equal letters, other edges to edit operations.

longest common subsequences (see Figure 12.3). We denote by $lcs(x, y)$ the maximal length of subsequences common to x and y. For fixed x and y, we denote by $LCS[i, j]$ the length of a longest common subsequence of $x[1 . . i]$ and $y[1 . . j]$.

There is a strong relationship between the longest common subsequence problem and a particular edit distance computation. Let $edit_{di}(x, y)$ be the minimal number of operations *delete* and *insert* necessary to transform x into y. This corresponds to a restricted edit distance where changes are not allowed. A change in the edit distance can be replaced by a pair of operations, deletion and insertion. The following lemma shows that the computation of $lcs(x, y)$ is equivalent to the evaluation of $edit_{di}(x, y)$. The statement is not true in general for all edit operations, or for edit operations having distinct costs. However, the restricted edit distance $edit_{di}$ remains to give weight 2 to changes, and weight 1 to both deletions and insertions. Recall that x and y have respective length m and n, and let $EDIT_{di}[i, j]$ be $edit_{di}(x[1 . . i], [1 . . j])$.

Lemma 12.1 *We have* $2 \cdot lcs(x, y) = m + n - edit_{di}(x, y)$, *and* $2 \cdot LCS[i, j] = i + j - EDlT_{di}[i, j]$, *for* $0 \le i \le m$ *and* $0 \le j \le n$.

Proof. The equation can be easily proved by induction on i and j. This is also apparent on the graphical representation in Figure 12.4. Consider a path from the source to the sink in which diagonal edges are only allowed when the corresponding symbols are equal. The number of diagonal edges in the path is the length of a subsequence common to x and y. Horizontal and vertical edges correspond to a sequence of edit operations (delete, insert) to transform x into y. If we assign cost 2 to the diagonal edges, the length of the path is exactly the sum of the lengths of words, $m + n$.   □

As a consequence of Lemma 12.1 computing $lcs(x, y)$ takes the same amount of time as computing the edit distance of the strings. A longest common subsequence can even be found with linear extra space (see bibliographic references).

Theorem 12.2 *A longest common subsequence of two strings of lengths* m, n *can be computed in* $O(mn)$ *time using* $O(mn)$ *additional memory.*

Proof. Assume that the table $EDIT_{di}$ is computed for zero-one costs of edges. Table LCS can be precomputed from Lemma 12.1. After that, a longest common subsequence can be constructed from the table LCS.   □

Let r be the number of shaded boxes in Figure 12.3. More formally, it is the number of pairs (i, j) such that $x[i] = y[j]$. If r is small (which happens often in practice) compared to mn, then there is an algorithm to compute longest common subsequence that is faster than the previous algorithm.

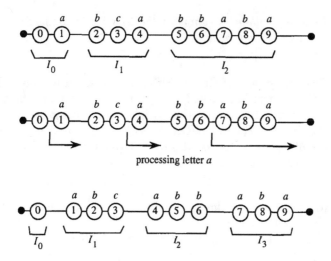

Figure 12.5: Hunt-Szymanski strategy on the word $y = abcabbaba$. Partitions of positions just before processing letter a of $x = cba$ (top), and just after (bottom).

The algorithm is given below. It processes the word x sequentially from left to right. Consider the situation in which $x[1 .. i-1]$ has just been processed. The algorithm maintains a partition of positions on the word y into intervals $I_0, I_1, \ldots, I_k, \ldots$ that are defined by

$$I_k = \{j : lcs(x[1 .. i-1], y[1 .. j]) = k\}.$$

In other words, positions in a class I_k correspond to prefixes of y having the same maximal length of common subsequence with $x[1 .. i-1]$. Consider, for instance, $y = abcabbaba$ and $x = cb \ldots$ Figure 12.5 (top) shows the partition $\{I_0, I_1, I_2\}$ of positions on y. For the next symbol of x, letter a, the figure (bottom) displays the modified partition $\{I_0, I_1, I_2, I_3\}$. The computation reduces to shifting to the right, like bowls on a wire, positions corresponding to occurrences of a inside y.

The algorithm lcs below implements this strategy. It makes use of operations on intervals of positions: *CLASS, SPLIT*, and *UNION*. They are defined as follows. For a position p on y, $CLASS(p)$ is the index k of the interval I_k to which it belongs. When p is in the interval $[f, f+1, \ldots, g]$, and $p \neq f$, then $SPLIT(I_k, p)$ is the pair of intervals $([f, f+1, \ldots, p-1], [p, p+1, \ldots, g])$. Finally, *UNION* is the union of two intervals; in the algorithm, only unions of disjoint consecutive intervals are performed.

Theorem 12.3 *Algorithm lcs computes the length of a longest common subsequence of words of length m and n ($m < n$) in $O((n + r) \log n)$ time, where $r = |\{(i,j) : x[i] = y[j]\}|$.*

Proof. The correctness of the algorithm is left as an exercise. The time complexity of the algorithm strongly relies on an efficient implementation of intervals I_k's. Using an implementation with B-trees, it can be shown that each operation $CLASS$, $SPLIT$, and $UNION$ takes $O(\log n)$ time. Preprocessing lists of occurrences of letters of the word y takes $O(n \log n)$ time. The rest of the algorithm takes $O(r \log n)$ time.   □

```
function lcs(x, y): integer; { Hunt-Szymanski algorithm }
{ m = |x| and n = |y| }
    I_0 := {0, 1, ..., n}; for k := 1 to n do I_k = ∅;
    for i := 1 to m do
        for each p position of x[i] inside y
                in decreasing order do begin
            k := CLASS(p);
            if k = CLASS(p − 1) then begin
                (I_k, X) := SPLIT(I_k, p);
                I_{k+1} := UNION(X, I_{k+1});
            end;
        end;
    return CLASS(n);
```

According to Theorem 12.3, if r is small, the computation of *lcs* by the last algorithm takes $O(n \log n)$ time, which is faster than with the dynamic programming algorithm. But, r can be of order mn (in the trivial case where $x = a^m, y = a^n$, for example), and then the time complexity becomes $O(mn \log n)$, which is larger than the running time of the dynamic programming algorithm.

The problem of computing *lcs* can be reduced to the computation of the longest increasing subsequence of a given string of elements belonging to a linearly ordered set. Let us write the coordinates of shaded boxes (as in Figure 12.3) from the first to the last row, and from left to right within rows. By doing so, we get a string w. No matrix table is needed to build w, the words x and y themselves suffice. For example, for the words of Figure 12.4 we get the sequence

$$((1,3), (2,2), (2,5), (2,6), (3,1), (3,4), (3,7),$$
$$(4,2), (4,5), (4,6), (5,1), (5,4), (5,7), (6,3)).$$

Define the following linear order on pairs of positions:

$$(i, j) \ll (k, l) \text{ iff } ((i = k) \text{ and } (j > l)) \text{ or } ((i < k) \text{ and } (j < l)).$$

Then a longest increasing (according to $\ll$) subsequence of the string w gives a longest common subsequence of the words x and y. There is an elegant algorithm to compute such increasing subsequences running is $O(r \log r)$ time. It presents an alternative to the above algorithm.

12.3 String matching with errors

String matching with errors differs only slightly from the edit distance problem. Here, we are given pattern *pat* and text *text* and we want to compute $\min(edit(pat, y) : y \in \mathcal{F}(text))$. Simultaneously, we want to find a factor y of *text* realizing the minimum and the position of one of its occurrences on *text*. We consider the table SE (that stands for String matching with Errors), of the same type as table $EDIT$:

$$SE[i, j] = \min(edit(pat[1..i], y) : y \in \mathcal{F}(text[1..j])).$$

The computation of table SE can be executed with dynamic programming. It is very similar to the computation of table $EDIT$.

Theorem 12.4 *The problem of string matching with errors can be solved in* $O(mn)$ *time.*

Proof. Surprisingly the algorithm is almost the same as that for computing edit distances. The only difference is that we initialize $SE[0, j]$ to 0, instead of j for $EDIT$. This is because the empty prefix of *pat* matches an empty factor of *text* (no error). The formula $(*)$ also works for SE. Then, $SE[m, n]$ is the distance between *pat* and one of its best matches y in the text. To find an occurrence of y and its position in *text*, we can use the same graph-theoretic approach as we used for the computation of longest common subsequences. It is recovered by a trace back inside the computed table SE from an extremal value. This completes the proof. □

One of the most interesting problems related to string matching with errors concerns the case in which the allowed number of errors is bound by a constant k. The number k is usually understood as a small fixed constant. We show that this problem can be solved in $O(n)$ time, or more exactly in $O(kn)$ time, if k is not fixed. For a fixed value of the parameter k, this gives an algorithm

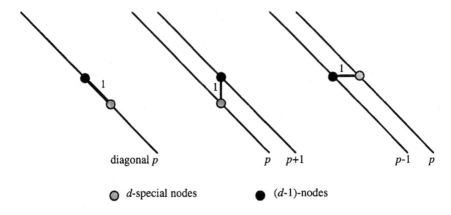

Figure 12.6: The computation of d-special nodes: nodes reachable by one edge of weight 1 from a $(d-1)$-node.

having an optimal asymptotic time complexity. Recall that the string edit table SE is computed according to the recurrence:

(∗) $SE[i,j] = \min(SE[i-1,j]+1, SE[i,j-1]+1, SE[i-1,j-1]+\partial(x[i],y[j]))$,

for words x and y.

Suppose that we have a fixed bound k on the number of errors. We require that the complexity is $O(kn)$. Only $O(kn)$ entries of the table must be considered. The basic algorithmic trick is to consider only so-called d-nodes, which are entries of the table SE satisfying special conditions. The d-nodes are defined in such a way that altogether we have $O(kn)$ such nodes.

We consider diagonals of the table SE. Each diagonal is oriented top-down, left-to-right. We define a d-node as the last pair (i,j) on a given diagonal satisfying $SE[i,j] = d$. Note that it is possible that a diagonal has no d-node. The approximate string-matching problem reduces to the computation of d-nodes. And it is clear that there is an occurrence of the pattern with d errors ending at position j on $text$ iff (m,j) is a d-node.

Computation of d-nodes is executed for $d = 0, 1, \ldots, k$ in this order. Computation of 0-nodes is equivalent to string matching without errors. Assume that we have already computed the $(d-1)$-nodes. To compute d-nodes we need two auxiliary concepts: d-special nodes, and maximal subpaths of zero-weight on a given diagonal. For a node (i,j), define the node $NEXT(i,j) = (i+t, j+t)$ as the lowest node on the same diagonal as (i,j), reachable from (i,j) by a subpath of zero weight. The subpath can be of zero length, and in this case $NEXT(i,j) = (i,j)$. A d-special node is a node reachable from a $(d-1)$-node

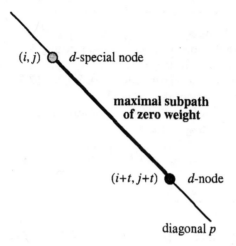

Figure 12.7: The computation of d-nodes from d-special nodes.

by one edge. Once d-special nodes are computed, the d-nodes can be easily found, as suggested in Figure 12.7. The structure of the algorithm is given below.

Theorem 12.5 *Assume that the alphabet is of a constant size. Approximate string matching with k errors can be achieved in $O(kn)$ time.*

Proof. It is sufficient to prove that we can run $O(kn)$ calls of the function *NEXT* in $O(kn)$ time. The equation $NEXT(i,j) = (i + t, j + t)$ means that t is the size of the longest common prefix of $pat[i..m]$ and $text[j..n]$. Assume that we have computed the common suffix tree for words pat and $text$. The computation of the longest common prefix of two suffixes is equivalent to the computation of their lowest common ancestor LCA in the tree. There is an algorithm (mentioned in Chapter 5) that preprocesses any tree in linear time in order to allow further LCA queries in constant time. This completes the proof. $\square$

Algorithm *Approximate string-matching with at most k errors*;
    compute 0-nodes { exact string-matching }
    for $d := 1$ **to** k **do begin**
        compute d-special nodes; { see Figure 12.6 }
        { computation of d-nodes }
        $S = \{ NEXT(i,j) : (i,j) \text{ is a } d\text{-special node} \}$;
        for each diagonal p **do**
            select on p, the lowest node that is in the set S;
            { selected nodes form the set of d-nodes }

12.4   String matching with don't care symbols

In this section, we assume that pattern *pat* and *text* can contain occurrences of the symbol $\oslash$, called the *don't care symbol*. Several different don't care symbols can be considered instead of only one, but the assumption is that they are all indistinguishable from the point of view of string matching. These symbols match any other symbol of the alphabet. We define an associated notion of *matching* on words as follows. We say that two symbols a, b match if they are equal, or if one of them is a don't care symbol (see Figure 12.8). We write $a \approx b$ in this case. We say that two strings u and v (of same length) match if $u[i] \approx v[i]$ for any position i. String matching with don't care symbols entails the problem of finding a factor of *text* that matches the pattern *pat* according to the present relation $\approx$.

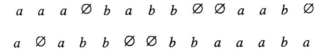

Figure 12.8: Two strings, with don't care symbols, that match.

String matching with don't care symbols does not use any of the techniques developed for other string-matching questions. This is because the relation $\approx$ is not transitive. Moreover, if symbol comparisons (involving only the relation $\approx$) are the only access to input texts, then there is a quadratic lower bound for the problem, which additionally proves that the problem is quite different from other string-matching problems. The algorithm presented later is an interesting example of a reduction of a textual problem to a problem in arithmetics.

Theorem 12.6 *If symbol comparisons are the only access to input texts, then $\Omega(n^2)$ such comparisons are necessary to solve the string-matching problem with don't care symbols.*

Proof. Consider a pattern of length m, and a text of length $n = 2m$, both consisting entirely of don't care symbols $\oslash$. Occurrences of the pattern start at all positions $1, \ldots, m$. If the comparison "$pat[j] \approx text[i]$," for $1 \le j \le m$ and $1 \le j \le n$, is not done, then we can replace $pat[j]$ and $text[i]$ by two distinct symbols a, b (that are not don't care symbols). The output then remains unchanged, but one of the occurrences is disqualified. Hence, the algorithm is not correct. This proves that all comparisons "$pat[j] \approx text[i]$" for $1 \le j \le m$ and $m < i \le n$, must be executed. $\qquad \square$

Contrary to what occurs elsewhere, we temporarily assume that positions in *pat* and *text* are numbered from zero (and not from one). We start with an algorithm that "multiplies" two words in a manner similar to how two binary numbers are multiplied, but ignoring the carry. We define the product operation $\bullet$ as the composition of $\approx$ and the logical "and" in the following sense. If x, y are two strings, then $z = x \bullet y$ is defined by

$$z[k] = \text{AND}(x[i] \approx y[j] : i + j = k),$$

for $k = 0, 1, \ldots, m + n - 2$. In other words, it is the logical "and" of all values $x[i] \approx y[j]$ taken over all i, j such that both $i + j = k$ and $x[i], y[j]$ are defined: We can write symbolically $\bullet = (\approx, \text{and})$.

Let p be the reverse of pattern *pat*, and consider $z = p \bullet text$. Let us examine the value of $z[k]$. We have $z[k] = $ true iff $(p[m-1] \approx text[k-m+1])$ and $p[m-2] \approx text[k-m+2]$ and $\ldots$ and $p[0] \approx text[k])$. Therefore, "$z[k] = $ true" exactly means that there is an occurrence of *pat* ending at position k in *text*. Hence, the string matching with don't care symbols reduces to the computation of the product $\bullet$.

Let us define an operation on logical vectors similar to $\bullet$. If x, y are two logical vectors, then $z = x \Diamond y$ is defined by

$$z[k] = \text{OR}(x[i] \ \text{and} \ y[j] : i + j = k).$$

For a word x and a symbol a, denote by $logical(a, x)$ the logical vector in which the i-th component is *true* iff $x[i] = a$. Define also

$$LOGICAL_{a,b}(x, y) = logical(a, x) \ \Diamond \ logical(b, y).$$

The following fact is now apparent: for two words x, y, the vector $x \bullet y$ equals the negation of logical OR of all vectors $LOGICAL_{a,b}(x, y)$ over all distinct symbols a, b that are not don't care symbols.

A consequence of the above fact is that, for a fixed-size alphabet, the complexity of evaluating the product • is of the same order as that of computing the operation ◊. Now, we show that the computation of the operation ◊ can be reduced to the computation of the ordinary product * of two integers. Let x, y be two logical vectors of size n. Let $k = \log n$. Replace logical values *true* and *false* by ones and zeros, respectively. Next, insert an additional group of k zeros between each two consecutive digits. We obtain binary representations of two numbers x', y'. Let z' be the integer $x * y$. The vector $z = x \diamond y$ can be recovered from z' as follows. Take the first digit (starting at position 0), and then each $(k + 1)$-th digit of z'; convert one and zero into *true* and *false*, respectively. In this way, we have proven the following statement.

Theorem 12.7 *The string-matching problem with don't care symbols can be solved $IM(n \log n)$ time, where $IM(r)$ denotes the complexity of multiplying two integers of size r.*

The value $IM(r)$ depends heavily on the model of computations considered. If bit operations are counted, then the best known algorithm for the problem is given by the Schönhage-Strassen multiplication, which works in time only slightly larger than $O(r \log r)$. No linear-time algorithm for the problem is known. This gives an $O(n \log^2 n)$-time algorithm for the string-matching problem with don't care symbols. String matching with don't care symbols generates a methodological interest because of its relationship to arithmetics. It would be also interesting to find relationship between some other typical textual problems to arithmetics.

Bibliographic notes

The edit distance computation can be attributed to Needleman and Wunsch [NW 70] and to Wagner and Fischer [WF 74]. Various applications of sequence comparisons are presented in a book edited by Sankoff and Kruskal [SK 83]. Approximate string-matching algorithms are widely used for molecular sequence comparisons, for which a large number of variants have been developed (see, of example [GG 89]). Computation of a longest common subsequence (not only its length) in linear space is from Hirschberg [Hi 75]. The last algorithm of Section 12.2 is from Hunt and Szymanski [HS 77]. It is the base of the "diff" command of UNIX system. An improvement on it is due to Apostolico and Guerra [AG 87].

An algorithm for the longest increasing subsequence can be found in [Ma 89].

There have been numerous substantial contributions to the problem, among them are those by Hirschberg [Hi 77], Nakatsu, Kambayashi, and Yajima [NKY 82], Hsu and Du [HD 84], Ukkonen [Uk 85b], Apostolico [Ap 85] and [Ap 87], Myers [My 86], Apostolico and Guerra [AG 87], Landau and Vishkin [LV 89], Apostolico, Browne, and Guerra [ABG 92], Ukkonen and Wood [UW 93].

A subquadratic solution (in $O(n^2/\log n)$ time) to the computation of edit distances has been given by Masek and Paterson [MP 80] for fixed-size alphabets. A subquadratic solution for unrestricted cost functions may be found in [CLU 02].

The first efficient string matching with errors is by Landau and Vishkin [LV 86b]. The computation of lowest common ancestors (LCA) is discussed by Schieber and Vishkin in [SV 88] (see the bibliographic notes at the end of Chapter 5). The best asymptotic time complexity of the string matching with don't care symbols is achieved by the algorithm of Fischer and Paterson [FP 74]. Practical approximate string matching is discussed by Baeza-Yates and Gonnet [BG 92], and by Wu and Manber [WM 92]. These solutions are close to each others. The second algorithm is implemented under UNIX as command "agrep."

Chapter 13

Matching by dueling and sampling

In this chapter we present a non-classical string-matching algorithm in which the preprocessing phase is closely related to borders of words and to *KMP* algorithm. We also introduce an interesting new operation called the *duel*. A more essential use of this operation can be seen in optimal parallel string matching and two-dimensional pattern matching. Hence, this section can be treated as a preparation for more advanced algorithms to be presented later.

13.1 String matching by duels

We assume in this section that the pattern *pat* is non-periodic, which means that its smallest period is larger than $|pat|/2$. This assumption implies that two consecutive occurrences of the pattern in a text (if any) are at a distance greater than $|pat|/2$. However, it is not clear how to use this property for searching the pattern. We proceed as follows: after a suitable preprocessing phase, given too close positions in the text, we eliminate one of them as a candidate for a match. This leads to the idea of a duel. The basic table which enables us to search for the pattern created by a duel-based algorithm can be computed either as a side effect of *KMP* algorithm, or by use of the table *Bord*. Duels are performed at search phase. Finally we define the following *witness table WIT*: for $0 < i < |m|$,

$$WIT[i] = \textbf{any } k \text{ such that } pat[i+k] \neq pat[k], \text{ or}$$
$$WIT[i] = 0, \text{ if there is no such } k.$$

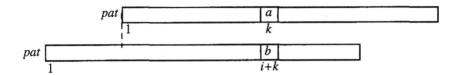

Figure 13.1: Witness of mismatch $a \neq b$, $k = WIT[i]$.

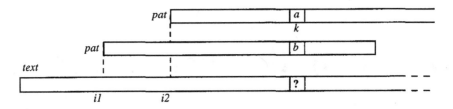

Figure 13.2: Duel between two inconsistent positions i_1 and i_2. One of them is eliminated by comparing symbol "?" in the text with a and b.

This definition is illustrated in Figure 13.1. A position i_1 on *text* is said to be *in the range of* a position i_2 if $|i_1 - i_2| < m$. We also say that two positions $i_1 < i_2$ on the text are *consistent* if i_2 is not in the range of i_1, or if $WIT[i_2 - i_1] = 0$. If the positions are not consistent, then we can remove one of them as a candidate for the starting position of an occurrence of the pattern just by considering position $i_2 + WIT[i_2 - i_1]$ on the text. This is the operation called a duel (see Figure 13.1). Let $i = i_2 - i_1$, and $k = WIT[i]$. Assume that we have $k > 0$, that is, positions i_1, i_2 are not consistent. Let $a = pat[k]$ and $b = pat[i + k]$, then $a \neq b$. Let c be the symbol in the text at position $i_2 + k$; it is indicated by "?" in Figure 13.1. We can eliminate at least one of the positions i_1 or i_2 as a candidate for a match by comparing c with a and b. In some situations, both positions can be eliminated, but, for simplicity, the algorithm below always removes exactly one position. Let us define, with $a = pat[WIT[i_2 - i_1]]$:

$$duel(i_1, i_2) = (\textbf{if } a = c \textbf{ then } i_2 \textbf{ else } i_1).$$

The position that "survives" is the value of *duel*, the other position is eliminated.

Assume the witness table is computed. It is then possible to reduce the search for *pat* in *text* to the search for pattern 1^m (repetition of m 1's) in a text of 0's and 1's. This last problem is obviously simpler than the general string-matching problem and can be solved in linear time (essentially with one counter). The following property of consistent positions (transitivity) is crucial for the correctness of the algorithm. It is called the *consistency property*:

let $i_1 < i_2 < i_3$; if i_1, i_2 are consistent and i_2, i_3 are consistent

then i_1, i_3 are also consistent.

Using this property we are able to eliminate a set of candidate positions from the text in such a way that all remaining positions are pairwise consistent. This can be done using the mechanism of stack (pushdown-store). Assume we have a stack of positions satisfying the property: positions are pairwise consistent, and in increasing order (from the top of the stack). Then, if we push onto the stack a position that is both smaller than the top position and consistent with it, the stack retains the property.

A set of consistent positions is *complete* if an occurrence of the pattern cannot start at any position that is not in the set. We say that a position x in the text agrees with a candidate position y if the symbol at position x agrees with the corresponding symbol of the pattern when it is placed at position y (that is, $text[x] = pat[x - y]$). Assume that S is a complete set of consistent positions, x is any position in the text, and y is any candidate position in S such that x is in the range of y with $x > y$. Then, as a consequence of the consistency property, we have the following equivalence:

x agrees with y iff x agrees with all positions in S.

Hence, it is sufficient to check the agreement of each position with any position from a set of consistent positions, see Figure 13.2. In this checking, we flag x with the value 0 or 1 depending on the agreement. Using this feature, the string matching reduces to the matching of unary patterns (patterns consisting entirely of ones).

The duel-based algorithm uses an additional zero-one vector, called *text1*. The value of the vector *text1* computed by the algorithm satisfies: pattern 1^m occurs at position i in *text1* iff the original pattern occurs at i in the text.

The algorithm obviously has linear-time complexity. Moreover, this complexity does not depend on the size of the alphabet. The basic component that remains to be shown is the computation of the witness table *WIT* on the pattern.

Later we shall see, in the case of parallel computations, that the fact that *any* position k for a witness is possible has a great importance. This is sufficient to choose any position, which is easier to compute in parallel. However, in the case of sequential computations, we can take the smallest such position as a witness. This leads us to define *WIT*[i] as *PREF*[i], for each position i such that $i + WIT[i] < m$ and $WIT[i] = 0$ for others. Section 3.2 contains both the definition of *PREF*, and a linear-time algorithm to compute it. The complexity of this latter algorithm is independent of the size of the alphabet, which shows the following statement.

Theorem 13.1 *String matching by duels takes linear time (search phase and pre-processing phase).*

We believe that matching by duels is one of the basic algorithms, since the idea of duels is the key to the optimal parallel string matching of Vishkin. Historically, however, the first optimal (for fixed alphabets) parallel string-matching algorithm uses another type of duels that we call *expensive duels*. Its advantage is that no additional table, like the one of witnesses, is needed. Its drawback is that the resulting algorithm is not optimal.

Let us observe the very parallel nature of this algorithm at a given stage k: the actions on all k-blocks can be performed simultaneously.

---

function *expensive-duel*(i, j) : integer;
   $k := \log_2(j - i) + 1$;
   if $text[i + 1..i + 2^k] = pat[1..2^k]$ **then return** i **else return** j;

Theorem 13.2 *Assume that all prefixes of pat in the form* $pat[1..2^k]$ *are non-periodic. Then, the algorithm String-searching-by-expensive-duels takes* $O(n \log m)$ *time.*

Proof. At stage k we consider only $O(n/2^k)$ survivals. Each expensive duel at this stage takes $O(2^k)$ time. There are $m/2$ stages. Together this gives $O(n \log m)$ time. This completes the proof. □

The expensive duels are restricted in use. Historically, however, they appeared before the concept of duel appeared. This is the only reason why it is reported here. The function *expensive-duel* is similar to the function *duel*, but its computation is much more expensive. This is the reason for the name "expensive duel." In the searching algorithm below, we partition the text into disjoint blocks of size 2^k. We call them k-blocks.

---

function *String-searching-by-duels*: Boolean;
{ Let S be a stack of positions }
   $S :=$ empty stack;
   for $i := n$ **down to** 1 **do begin**
      push i on the stack S;
      while $|S| > 2$ **and**
              the top two elements i_1, i_2 of S are inconsistent **do**
            replace them in S by the single element $duel(i_1, i_2)$;
   end;
   mark in the text all positions that are in S;
   { all marked positions are pairwise consistent }
   for $i := 1$ **to** n **do begin**
      $k :=$ first marked position to the left of i, including i;
      if k undefined **or** $pat[i - k + 1] \neq text[i]$ **then** $text1[i] := 0$
      else $text1[i] := 1$;
   end;
   if $text1$ contains the pattern 1^m **return** true
   else return false;

We only show the use of expensive duels for a special type of pattern. Assume that the size of the pattern is a power of two. Assume also that $pat[1..2^k]$ is non-periodic, for each $1 \leq k \leq \log n$. Here, such patterns are called special patterns.

Algorithm *String-searching-by-expensive-duels*;
{ assume $n - m + 1$ and m are powers of two }
{ assume that the pattern is special }
   initially all positions in $[1..n - m]$ are survivals;
   for $k := 1$ **to** $\log n$ **do**
      for each k-block **do begin**
         let i and j be the only survivals in the k-blocks;
         make $expensive\text{-}duel(i, j)$ a survival;
      end;
   { there are $O(n/m)$ survivals }
   for each survival position i **do**
      check occurrence of *pat* at position i naively;

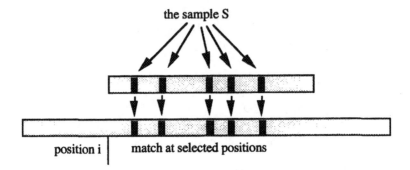

Figure 13.3: An occurrence of the sample S in the text.

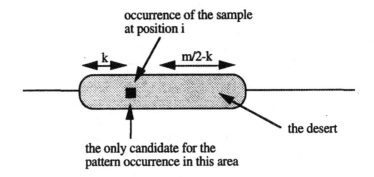

Figure 13.4: An occurrence of the sample, and the desert area.

13.2   String matching by sampling

Both *KMP* algorithm and *BM* algorithm (Chapter 2) scan symbols at consecutive positions on the pattern. The first one scans a prefix of the pattern, and the second one scans a suffix of the pattern. In this section, we show an algorithm that first scans a sequence of not necessarily consecutive positions on the pattern, and then, in case of success, completes the scan of the pattern. The first scanning sequence is called a *sample*.

A sample S for the pattern *pat* is a set of positions on *pat*. A sample S occurs at position i in the text if $pat[j] = text[i+j]$ for each j in S (see Figure 13.3).

A sample S is called a *good sample* if it satisfies the two conditions (see Figure 13.4):

1. S is small: $|S| = O(\log m)$,

2. there is an integer k such that if S occurs at position i in the text then no occurrence of the pattern starts in the segment $[i - k .. i + m/2 - k]$, except perhaps at position i. The segment is called the *desert*.

If the pattern has period p, setting $k = min(m, 2.p - 1)$, the prefix $pat[1..k]$ is called the non-periodic part of the pattern.

Theorem 13.3 *Assume we are given the period of the pattern pat, and a good sample of its non-periodic part, then, the search for pat can be done in $O(n \log n)$ time with only $O(\log m)$ additional memory space.*

Proof. Assume for a moment that the pattern itself is non-periodic. Let us partition the input text into windows of size $m/2$. We consider each window separately, and find the first and the last positions of occurrences of the sample in the window (if there are at least two occurrences). These occurrences only are possible candidates for an occurrence of the pattern. Each of these occurrences is checked in a naive way (constant size additional memory is sufficient for that). This proves that the non-periodic part of the pattern can be found in the text with the required complexity. The general case of periodic patterns is left as an exercise. One has to find sufficiently many consecutive occurrences of the non-periodic part of the pattern. An additional counter is needed to remember the number of consecutive occurrences. This completes the proof. □

Theorem 13.4 *If the pattern is non-periodic then it has a good sample S. The sample can be constructed in linear time.*

Proof. Assume we have computed the witness table *WIT* (see Section 13.1). Let us consider potential occurrences of the pattern at positions $1, 2, \ldots, m/2$ of some imaginary text. Let us identify these pattern occurrences with numbers $1, 2, \ldots, m/2$. The occurrence corresponding to the i-th position is called the i-th row. If we draw a vertical line at a position j, then it can intersect a given i-th row or not. If it intersects, then there is a symbol at the point of intersection (see Figure 13.5). Let us denote this symbol by $symbol(i, j)$.

Claim 1. Let i_1, i_2 be two different elements of $[1..m/2]$. Then, there is an integer j such that the j-th column intersects both rows i_1 and i_2; Moreover, $symbol(i_1, j) \neq symbol(i_2, j)$. The integer j can be found in constant time if the witness table of the pattern is precomputed.

The claim is a reformulation of the property of non-periodicity. Due to non-periodicity, for occurrences of the pattern placed at positions i_1, i_2 there is a

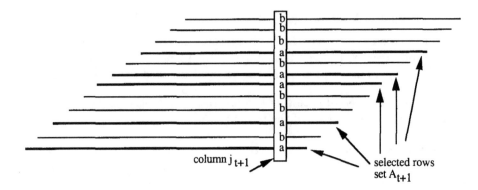

Figure 13.5: The set $A_{t+1} = \{i \in A_t : Symbol(i, j_{t+1}) = a\}$ is the smaller one.

mismatch position j given by $j = i_2 + WIT[i_2 - i_1]$. This means that if we look at the column placed at position j, then this column intersects occurrences of the pattern with different symbols.

Claim 2. Let J be any set of rows. If a vertical column intersects the first and the last row of J, then it intersects all the rows of J.

We now prove an equivalent geometrical formulation of the thesis of the theorem.

Claim 3. There is a row i and a set J of $O(\log m)$ vertical columns placed at positions $j_1, j_2, \ldots, j_k$ such that:

1. all columns in J intersect the row i,

2. if $i \neq r$ ($r \in [1..m/2]$), then there is a column j in J intersecting rows i and r such that $symbol(i, j) \neq symbol(r, j)$.

Proof of the claim. We construct the set J and the row i by the algorithm below, that ends the proof of the theorem. $\qquad\square$

Algorithm *Find-good-sample*;
    $J :=$ empty set; $A_0 := [1..m/2]$; $t := 0$;
    while $|A_t| > 1$ **do begin**
        find any column j_{t+1} that intersect all rows of A_t
        with two different symbols at intersection points;
        { use Claim 1 and Claim 2 }
        Let a, b be the two different symbols at intersections;
        $A_{t+1} :=$ smaller of the two sets:
            $\{i \in A_t : symbol(i, j_{t+1}) = a\}$
            and $\{i \in A_t : symbol(i, j_{t+1}) = b\}$;
        add j_{t+1} to J; $t := t + 1$;
    end;
    let i be the unique element of A_t;
    return(J, i);

Bibliographic notes

The ideas of duels and samples can be attributed to Vishkin who applied it to the design of parallel algorithms [Vi 85], [Vi 90] (see Chapter 16). Expensive duels were implicitly considered by Galil in [Ga 85].

Chapter 14

Two-dimensional pattern matching

The two-dimensional pattern matching is interesting due to its relationship to image processing. The efficiency of algorithms is even more important in the two-dimensional case because the size of the problem, the number of pixels of images, is very large in practical situations.

We mainly consider rectangular images. The pattern-matching problem is to locate an $m \times m'$ pattern array PAT inside an $n \times n'$ (text) array T. The position of an occurrence of PAT in T, (see Figure 14.1), is a pair (i, j), such that

$$PAT = T[i+1..i+m, j+1..j+m'].$$

We present two different solutions to the two-dimensional pattern matching. The first reduces the problem to multi-pattern matching. The second is based on two-dimensional periodicities and the notion of duels; it is presented in the next chapter. The linear-time algorithms with constant coefficients independent on the size of the alphabet are presented in Chapter 15. We also consider non-rectangular patterns in relation to approximate matching.

The method of sampling is presented here for two-dimensional patterns and appears to be very powerful for almost all patterns.

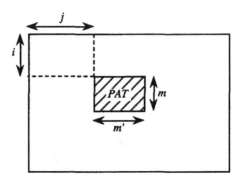

Figure 14.1: The pattern PAT occurs at position (i, j) in T.

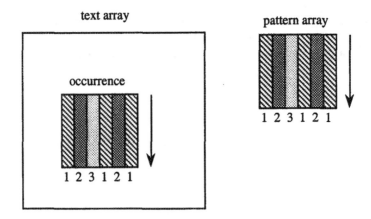

Figure 14.2: Two-dimensional pattern matching by searching for columns of the pattern.

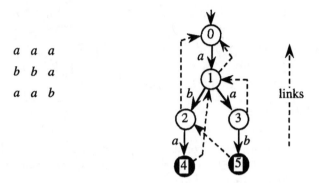

Figure 14.3: A pattern array, and the SMA automaton of its columns. Columns 1 and 2 are identified with state 4, column with state 5.

a	b	a	b	a	b	b		1	0	1	0	1	0	0
a	a	a	a	b	b	b		3	1	3	1	2	0	0
b	b	b	a	a	a	b		5	2	5	3	4	1	0
a	a	a	b	b	a	a		4	4	4	5	2	3	1
b	b	a	a	a	b	b		2	2	3	4	4	5	2
a	a	b	a	a	a	a		4	4	5	3	3	4	4

Figure 14.4: A text array, and its associated array of states (according to the SMA automaton of Figure 14.3).

14.1   Multi-pattern approach

The first solution to two-dimensional pattern matching is to translate it into a string-matching problem. The pattern is viewed as a set of strings, its columns. To locate columns of the pattern within columns of the text array requires searching for several patterns (see Figure 14.2). Moreover, the occurrences of patterns must be found in a particular configuration within rows: all columns of the patterns are to be found in the order specified by the pattern, and all ending on a same row of the text array. In this section, we use the Aho-Corasick approach (see Chapter 11) to solve the multi-pattern matching problem.

The strategy for searching for PAT in the text array T is as follows. Let Π be the set of all (distinct) columns of PAT (treated as words). We first build the string-matching automaton $G = SMA(\Pi)$ with terminal states (see Chapter

```
1  0  1  0  1  0  0          a  b  a  b  a  b  b
3  1  3  1  2  0  0          a  a  a  a  b  b  b
5  2  5  3  4  1  0          b  b  b  a  a  a  b
4  4  4  5  2  3  1          a  a  a  b  b  a  a
2  2  3  4  4  5  2          b  b  a  a  a  b  b
4  4  5  3  3  4  4          a  a  b  a  a  a  a
```

Figure 14.5: The pattern "445" corresponds to the pattern array of Figure 14.3 (left). Occurrences of "445" in the array of states give occurrences of the pattern array in the original text array.

11). Each terminal state corresponds to a pattern of Π. Therefore, columns of the patterns are identified with states of the SMA automaton. There can be less than m terminal states because of possible equalities between columns. Then, the automaton is applied to each column of T. We generate an array T' of the same size as T, and in which the entries are states of G. The pattern PAT itself is replaced by a string pat drawn from the set of states: the i-th symbol of pat is the state identified with the i-th column of PAT. The remainder of the procedure consists in locating pat inside the lines of T'. The strategy is illustrated by Figures 14.3, 14.4 and 14.5. This yields the subsequent result.

Theorem 14.1 *The two-dimensional pattern matching can be solved in time $O(N \log |A|)$, where $N = n \cdot n'$ is the size of the text array, and A is the alphabet.*

Proof. The time to build the automaton $SMA(\Pi)$ is $O(M \log |A|)$ where $M = m \times m'$ is the size of the pattern PAT. The construction of the array of column numbers T' takes $O(N \log |A|)$ time. The final search phase, string matching inside lines of T', takes $O(N)$ time.   □

The above algorithm seems to be inherently dependent upon the alphabet. This is because of the automaton approach. The existence of a linear-time alphabet-independent algorithm is discussed in the next chapter.

14.2   Don't cares and non-rectangular patterns

Assume that the two-dimensional pattern contains a certain number of holes. Holes can be regarded as filled with a special symbol that matches any other symbol. It is the don't care symbol $\oslash$ considered in Chapter 12 for approximate

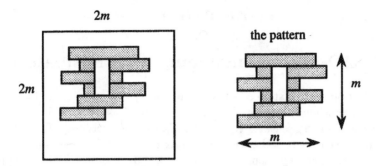

Figure 14.6: Searching a non-rectangular $m \times m$ pattern within pieces of shape $2m \times 2m$. $Lin(PAT)$ (see proof of Theorem 14.2) is independent of the position where PAT is placed inside shape S.

string matching. If the pattern is not rectangular, we can also complete it, adding enough don't care symbols so that it fits into an $m \times m'$ rectangle. By doing so, both questions become similar.

Theorem 14.2 *Two-dimensional pattern matching with don't care symbols, and pattern matching of non-rectangular patterns can be done in $O(N \log^2 m)$ time (with an $m \times m$ pattern, $m \geq m$, and an $n \times n'$ text array, $N = nn'$).*

Proof. We linearize the problem. Let PAT be a non-rectangular pattern that fits into an $m \times m$ rectangle, with $m \geq m$. We consider windows of shape $2m \times 2m$ on the text array (see Figure 14.6). We first solve the problem as if $n = 2m$. We define $Lin(PAT)$ as a one-dimensional version of PAT. It is a string with don't care symbols of $O(m)$ size constructed as follows: place PAT inside a $2m \times 2m$ shape S. All positions not occupied by PAT are filled with the don't care symbol $\oslash$; then, concatenate the rows of S, starting from the topmost row; within the string obtained in this way, remove the longest prefix and the longest suffix containing only don't care symbols. The resulting string is $Lin(PAT)$.

The basic property of the transformation Lin is: Let T be an $2m \times 2m$ text array. Let $Lin(T)$ be the string obtained by concatenating all rows of T, starting from the topmost row; then, searching for PAT in T is equivalent to searching for $Lin(PAT)$ inside $Lin(T)$. This can be executed using methods for string matching with don't care symbols (see Chapter 12). Then it is proven how to do it in $O(n \log^2 n)$ time, which here becomes $O(m^2 \log^2 m)$. A text array of size greater than $2m \times 2m$ can be decomposed into such (overlapping) sub-arrays on which the above procedure is applied. The total time becomes then $O(N \log^2 m)$. This completes the proof.   □

14.3   2D-Pattern matching with mismatches

The definition of a distance between two arrays is more complicated than
for (one-dimensional) strings. Insertions or deletions of a symbol can result
in an increase or decrease of the length of one row (column). Therefore, for
simplicity, we concentrate here on the approximate pattern matching with only
one edit operation: replacement of one symbol by another. This corresponds
to unit-cost mismatches.

For two strings x, y denote by $MISM_k(x, y, i)$ the set of k first (left-to-right)
mismatch positions between $y[i..i + |x| - 1]$ and x. We are not interested in
more than k mismatches.

Lemma 14.1 *Assume we are given two strings x and y, and their suffix tree
processed for LCA queries. Then, the computation of $MISM_k(x, y, i)$ can be
executed in $O(k)$ time for each position i in the text y.*

Proof. First we find the longest common prefix of $y[i..n]$ and x. This is
done using an *LCA* query for the leaves corresponding to y and x in the joint
suffix tree for these both texts. In this way, we obtain the first mismatch
position i_1.

Then, we look for the longest common prefix of $x[i_1 - i + 1..m]$ and $y[i_1..n]$.
This is again done by asking a suitable *LCA* query about leaves related to
$x[i_1 - i + 1..m]$ and $y[i_1..n]$. We obtain the next mismatch position (if its
exists). We continue in this way until k mismatch positions are found, or (in
the case where there are less than k mismatch positions) all mismatch positions
are found.

The time is proportional to the number of *LCA* queries, that is $O(k)$. This
completes the proof.   □

Theorem 14.3 *Assume the alphabet is of constant size. The problem of
matching with a fixed number k of mismatches a pattern within an $n \times n$ text
array can be solved in $O(kn^2)$ time.*

Proof. Let *PAT* be the $m \times m$ pattern, where $m \leq n$. The algorithm starts
as in the exact two-dimensional pattern matching, by a multi-pattern string
matching. The Aho-Corasick automaton for all columns of the pattern is
built. Then, the automaton is applied to all columns of the text T to obtain a
state array T'. The pattern array is replaced by a string of states *pat*.

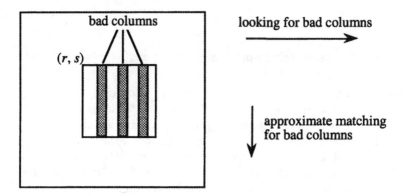

Figure 14.7: Approximate matching. There are at most k bad columns. If there is a match with at most k mismatches, then the total number of mismatches in all bad columns cannot exceed k.

Figure 14.7 illustrates how we check the approximate match at position (r, s) with at most k mismatches. Let y be the r-th row of T'. We compute $MISM_k(pat, y, s)$. This produces all columns that contain at least one mismatch with the pattern PAT placed at (r, s). Let us call these columns the bad columns.

We compute the total number of mismatch positions in bad columns with respect to the corresponding columns of the pattern (assuming it is placed at position (r, s)). We are only interested in a total of at most k mismatches. All mismatches are found by using the function $MISM$. The total complexity is proportional to the number of all LCA queries executed in the algorithm. We make at most k such queries.

Hence, for a fixed position (r, s) after the preprocessing, the complexity is $O(k)$. Since there is a quadratic number of positions, the total time complexity is as required.   □

14.4 Multi-pattern matching

In this section we consider a set of k square pattern arrays $X_1, X_2, \ldots, X_k$. For a given $n \times n$ text array T we want to check if any of these patterns occurs in T. This is the multi-pattern matching in two dimensions. Assume, for simplicity, that the size of the alphabet is constant. The strategy developed for Karp-Miller-Rosenberg algorithm (see Chapter 7) yields a solution to the general multi-pattern problem that works in $O(n^2 \log n)$ time. We omit the obvious proof.

Fact. Two-dimensional multi-pattern matching can be solved in $O(n^2 \log n)$ time using the algorithm *KMR*.

Indeed, the above result can be improved to $O(n^2 \log k)$ time, where k is the number of patterns. Of course, k can be of the same order as n, and this does not provide a substantial improvement. But we are also interested in alternative algorithms and some interesting new ideas behind them that enrich the algorithmics of two-dimensional matching. The natural alternative algorithm considered here is based on an extension of the Aho-Corasick string-matching automaton to the two-dimensional case. By the way, this also shows the important extension of the notion of *border*, *suffix*, and *prefix* to two-dimensional arrays. It also provides another pattern-matching algorithm for one pattern: it shows that searching the pattern along a fixed diagonal of the text array is reducible to one-dimensional string matching. Again, *LCA* preprocessing is crucial for the two-dimensional pattern-matching algorithm of the section. First consider the simple case in which all the patterns are of the same shape. Assume that they are $m \times m$ arrays. The method of Section 14.1 generally facilitates this situation. This gives a linear-time algorithm when all patterns are of the same size.

Theorem 14.4 *Two-dimensional multi-pattern matching can be solved in time $O(N)$ when the alphabet is fixed and all patterns are of the same size (where N is the total size of the problem).*

Proof. The algorithm works as follows. The Aho-Corasick machine is constructed for all columns of all patterns. Each pattern array is then transformed into a string of states. We obtain a set of strings $x_1, x_2, \ldots, x_k$. The text array T is replaced by the state array T' in the same way as in Section 14.1. Any multi-pattern string-matching algorithm then gives a solution. This gives a linear-time algorithm for this special case (fixed alphabet).   □

Next, we consider the general case, in which the patterns are square arrays of possibly different sizes. Again, the algorithm is an extension of the Aho-Corasick multi-pattern matching.

A prefix (resp. suffix) of a square array is a square sub-array containing the left top corner (resp. right bottom corner) of the array. We construct a two-dimensional version of the Aho-Corasick multi-pattern automaton $\mathcal{A}$ as follows. Each pattern is considered as a string: its i-th letter is the i-th segment of the array. The i-th segment is composed of the upper part of the i-th column, and the left part of the i-th row, beginning both at the i-th position on the diagonal (see Figure 14.8). The states of $\mathcal{A}$ are prefixes of all

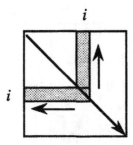

Figure 14.8: The i-th segment of a pattern array.

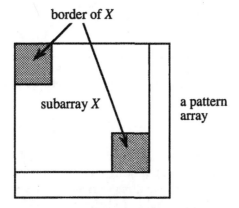

Figure 14.9: A border of a sub-array X of a pattern.

the pattern arrays. The set of states is organized in a tree in which the nodes correspond to the two-dimensional prefixes of the pattern.

The edges outgoing a node at depth $i - 1$ are labeled by the names of the i-th segments of the patterns. We can give consistent names to i-th segments of all patterns in time $O(k \log k)$ for a given i, since there are at most k such segments, one for each pattern. The equality of two segments can be checked in constant time using an *LCPref* query (a longest common prefix query) after a suitable preprocessing of the tree in which the edge labels are names of segments.

After that, the failure table *Bord* on the tree is built. The notion corresponds to borders of square arrays as illustrated in Figure 14.9. They are the largest proper sub-arrays that are both prefix and suffix of the given array.

We say that a segment π_1 is a part of a segment π_2 if the columns part of

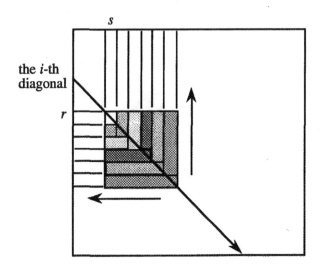

Figure 14.10: Checking the occurrences of a pattern at position (r, s) in the text array.

π_1 is a prefix of the column part of π_2 and a similar relation holds between rows of the segments. Define the following relation $==$ between two segments. Let π_1, π_2 be i-th and j-th segments, respectively, with $i \leq j$. Then, we write $\pi_1 == \pi_2$ iff either, both $i = j$ holds and the names of the segments are the same, or, π_1 is a part of π_2. The table *Bord* for the two-dimensional case is defined as for strings, except that relation $==$ is considered instead of equality.

Theorem 14.5 *Assume that the alphabet is fixed. Then the two-dimensional multi-pattern matching can be solved in time $O(N \log k)$ time, where k is the number of patterns.*

Proof. The two-dimensional pattern matching is essentially reduced to one-dimensional multi-pattern matching (see Figure 14.10). The equality of symbols is replaced by the relation $==$, and the corresponding table *Bord* works similarly.   □

14.5   Matching by sampling

The concept of *deterministic sample* introduced in Chapter 13 for one-dimensional patterns is very powerful. Its wide applicability appears, for example, in

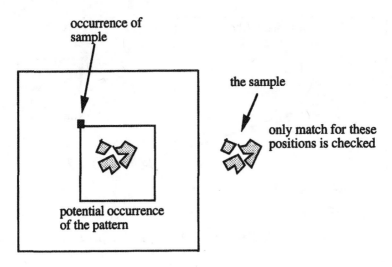

Figure 14.11: A deterministic sample S.

the domain of parallel computation, leading to a constant-time parallel string matching. The aim of this section is to extend and use the concept of a deterministic sample to the two-dimensional case. Almost every non-periodic pattern has a deterministic sample for which the properties are analogous to those of one-dimensional patterns. The use of 2-D sampling gives solutions both to sequential computation requiring only small extra space, and to constant-time parallel computation for the two-dimensional pattern-matching problem.

A *deterministic sample* S for PAT is a set of positions in the pattern PAT satisfying certain conditions. The sample S occurs at position $x = (i, j)$ in the text array if $PAT[y] = T[x + y]$ for each y in S (see Figure 14.11).

The central idea related to samples is the *field of fire* of the sample S similar to the *desert area* for strings. Finding an occurrence of the sample in the text assures us that there is an $m/2 \times m/2$ square in the text, called the field of fire of the occurrence of S, where there is only one possible matching position of the entire pattern. This possible matching position is $z = (k, l)$ relative to the origin of the field of fire (see Figure 14.12). Let x be a position of S in the text array. Let us denote by $fofire(x, S)$ its corresponding field of fire: it is the $m/2 \times m/2$ sub-square of the text array at position $x - z$. The field of fire should satisfy the *field of fire condition* described below.

Whenever the sample S occurs at position x in the text array, then there is no occurrence of the pattern within the area $fofire(x, S)$ of the text array, except maybe at position x. The size of a square S, denoted by $|S|$, is defined

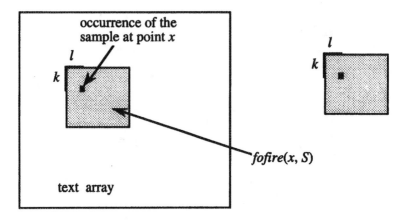

Figure 14.12: The field of fire of the sample S.

here as the length of its side. The deterministic sample S must be small but effective in "killing" other positions. It must satisfy the following conditions:

(*) $|S| = O(\log m)$,

(**) $|fofire(x, S)| \geq m/4$.

We consider only very particular samples. They are special segments: horizontal factor of length $4 \log m$ at position $(m/2, m/2)$ in the pattern (see Figure 14.13). Moreover, we say that the pattern PAT is *good* if its special segment occurs only once in PAT. Note that any segment lying "far away enough" from the boundaries of the array would work as well.

Theorem 14.6 *Assume that the alphabet contains at least two symbols. Then:*

(a) *almost all patterns are good,*

(b) *for almost all patterns there is a logarithmic-size sample for which the field of fire is an $m/2 \times m/2$ square,*

(c) *both, the sample can be found, and the goodness of the pattern can be checked in constant extra space and linear time.*

Proof. The first point follows by simple calculations. If the patterns is good the special segment is the sample. It is of logarithmic size. Its field of fire is the left upper $m/2 \times m/2$ quadrant of the pattern. If the sample occurs at position x in the text array, then no occurrence of the pattern in the text has position

$x + (k, l)$ for $0 \leq k \leq m/2$, $0 \leq l \leq m/2$, and $(k, l) \neq (0, 0)$. In that case there would be two occurrences of the special segment in the pattern. The last point follows from the fact that all occurrences of the special segment can be found within claimed complexities using algorithms for string matching. This completes the sketch of the proof. □

Theorem 14.7 *Assume that the alphabet contains at least two symbols. Then the two-dimensional pattern matching can be solved in constant extra space and linear time, for almost all patterns,*

Sketch of the proof. Only good patterns and considered, so results hold only for them. The whole text array is partitioned into $m/2 \times m/2$ windows. Within each window, we search for occurrences of the special segment with a serial algorithm working with the claimed complexities. This is done for all windows independently, window after window.

14.6 An algorithm fast on the average

A natural problem related to pattern matching is designing algorithms that are fast in practice. Since the notion of "practice" is not well defined, it is often considered for algorithms that are fast on the average. In this section, we design an algorithm making $O((N \log M)/M)$ comparisons on the average for the two-dimensional pattern matching (the pattern is an $m \times m$ array, the text is an $n \times n$ array, $N = n^2$, and $M = m^2$). All symbols appear with the same probability independently of each other in arrays. If M is of the same order as N, the algorithm makes only $O(\log N)$ comparisons on the average. The method described here is similar to the use of special segments in Section 14.5.

For simplicity we assume that the alphabet has only two elements, and that each of the two symbols of the text is chosen independently with the same probability. Let r be equal to $4 \log m$.

The algorithm is similar to the algorithm *fast-on-average* presented at the end of Chapter 2 as a variation of Boyer-Moore algorithm for string matching.

Informal description of the algorithm.

1. Partition the text array into windows of shape $m \times m$; the sub-window of a window consists of the last r positions of the lowest row of the window;

2. first check if the text contained in the sub-window is a factor of any row of the pattern;

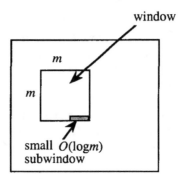

Figure 14.13: Searching for an occurrence of the pattern starting in the window; checking the sub-window first.

3. if so, search for an occurrence of the pattern having its left upper corner position in the window by any linear-time algorithm; apply the same procedure to each window.

The suffix of length r of the last row of the pattern behaves like a fingerprint. It has only logarithmic size, by definition. But it is unlikely to appear in subwindows. The test at line 2 above can be done with the help of a suffix tree or a suffix DAWG. On fixed alphabets, this takes $O(r)$ time. There are N/M windows, and a simple calculation shows the following (see also end of Chapter 2).

Theorem 14.8 *The two-dimensional pattern matching can be solved by doing $O(n^2 \log(m)/m^2)$ comparisons on the average, for fixed alphabets, after preprocessing the pattern.*

Bibliographic notes

The simple linear-time (on fixed alphabets) two-dimensional pattern-matching algorithms of Section 14.1 have designed independently by Bird [Bi 77] and Baker [Ba 78].

The algorithm to search for non-rectangular patterns is from Amir and Farach [AF 91].

The powerful concept of a deterministic sample for strings was introduced by Vishkin in [Vi 91]. The wide applicability of this concept was recently shown by Galil [Ga 92] who designed a constant parallel-time string matching (with a linear number of processors). The 2D-sampling method is from [CGR 92]. The algorithm "fast on the average" is a simple application of the similar algorithm described in Chapter 2. The idea comes from [KMP 77]; see also [BR 90].

The notion of a suffix tree on arrays is discussed by Giancarlo in [Gi 93].

Chapter 15

Two-dimensional periodicities

Similarly as in the one-dimensional case, the most interesting algorithms for exact two-dimensional matching are related to periodicities. However the structure of 2D-periodicities is much more complex, in particular the periodicity lemma (Section 1.7) is no longer directly applicable. For simplicity we assume that the pattern is an $m \times m$ square array of symbols.

A *period* of the pattern PAT is a non-null vector $p = (r, s)$ such that $-m < r < m$, $0 \le s < m$, and

$$PAT[i, j] = PAT[r + i, s + j]$$

whenever both sides of the equation are defined. Note that the second component of a period is assumed to be a non-negative integer, because we consider that period vectors are always oriented from left to right. There are two categories of periods, see Figure 15.1 and Figure 15.2, according to whether r is negative or not.

If there are close occurrences of the pattern in a text array, then there is an overlap of the pattern over itself, that is, a periodicity. If x and y are close positions of two occurrences PAT in the array T, assuming that y is to the right of x, the vector $y - x$ is a period of the pattern.

15.1 Amir-Benson-Farach algorithm

The algorithm of the present section is based on the idea of *duels*. The string-matching algorithm by duels presented in Chapter 13 for "one-dimensional"

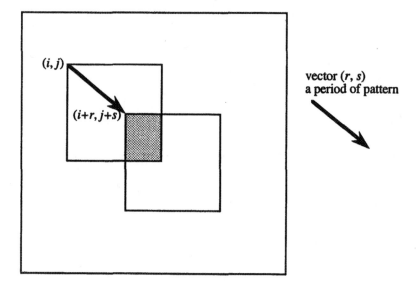

Figure 15.1: When $(i,j) <_t (k,l)$. If two occurrences of PAT overlap, PAT has a period $(r,s) = (k,l) - (i,j)$. Otherwise, a duel between (i,j) and (k,l) can be applied. to eliminate one possibility.

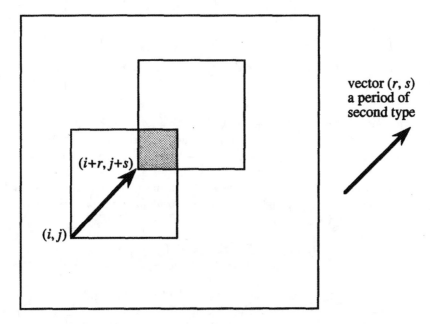

Figure 15.2: The second category of period, when $(i,j) <_b (k,l)$.

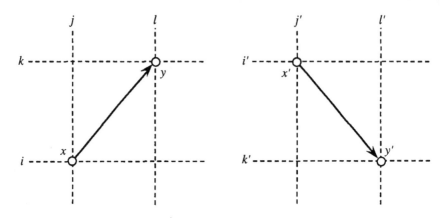

Figure 15.3: Ordering on positions: $x <_b y$ (left), $x' <_t y'$ (right).

strings extends to the two-dimensional case. The advantage of this approach is to produce a two-dimensional pattern-matching algorithm in which the search phase takes linear time, independently of the alphabet. The two-dimensional settings for duels, witnesses, and consistency relation are necessary for adapting the string-matching algorithm by duels. Positions in arrays are numbered top-down (rows) and left-to-right (columns). We define two partial orderings $<_b$ (for bottom) and $<_t$(for top) on positions on the array T:

$$(i,j) \leq_b (k,l) \text{ iff } i \geq k \text{ and } j \leq l,$$
$$(i,j) \leq_t (k,l) \text{ iff } i \leq k \text{ and } j \leq l.$$

The relation $x <_b y$ means that position x is to the left and at the *bottom* of y. The relation $x <_t y$ means that position x is to the left and at the *top* of y. For example, we have $(i,j) \leq_b (k,l)$ and $(i',j') \leq_t (k',l')$ in Figure 15.3. Making duels during the searching phase of the pattern-matching algorithm supposes, that we have an analogue to the *witness table* considered for strings. For arrays, the two-dimensional witness table WIT is defined as follows:

$$WIT[r,s] = \text{any position } (p,q) \text{ such that } PAT[p,q] \neq PAT[r+p,s+q],$$
$$WIT[r,s] = 0, \text{ if there is no such } (p,q).$$

The definition is illustrated in Figure 15.2 for the two categories of vector (r,s) (depending on whether $r \leq 0$ holds or not).

A duel is only performed on close positions according to the following notion. A position (k,l) is said to be *in the range of* the position (i,j) (according to the size of PAT) if $|k - i| < m$ and $|l - j| < m$. In addition, two positions x

and y such that y is to the right of x, are said to be *consistent* if y is not in the range of x, or if $WIT[y - x] = 0$, which means that $y - x$ is a period of PAT. Let us recall the notion of *duel*. If the positions x and y are not consistent, the pattern PAT cannot appear both at positions x and y inside the array T. In constant time, we can remove one of them as a candidate for a position of an occurrence of the pattern. Such an operation, called a *duel*, can be described as follows. Assume that positions x and y are not consistent, with y to the right of x. Let $z = WIT[y - x]$, $a = PAT[z]$ and $b = PAT[y - x + z]$. By definition of the witness table WIT, symbols a and b are distinct. Let c be the symbol $T[y + z]$. This symbol cannot be both equal to a and to b, so at least one of the positions x, y is not a matching position for PAT. If $b \neq c$, the pattern cannot occur at position x. If $a \neq c$, the same holds for y. Therefore, comparing c with a and b permits us to eliminate (at least) one of the positions. Note that in some situations both positions could be eliminated, however, for simplicity of the algorithm, only one position is always removed at a time. This is a mere duplication of the strategy developed for "one-dimensional" string matching. Let *duel* be defined by

$$duel(x, y) = (\text{if } b = c \text{ then } x \text{ else } y).$$

The value $duel(x, y)$ is the position that "survives", the other position is eliminated.

We now describe the two-dimensional pattern-matching based on duels. Assume the witness table of the pattern PAT is computed. Its precomputation is sketched at the end of this section. The first step of the searching phase reduces the problem to a two-dimensional pattern matching for unary patterns, as if all entries of PAT were the unique symbol a. We want to eliminate a set of candidate positions from the text array T in such a way that all remaining positions are pairwise consistent. Removed positions cannot be matching positions of the pattern. Then, with each position x on the text array we associate the value 1 iff, after duels, it corresponds to the symbol compatible with occurrences of the pattern placed at any position in the range of x. Otherwise, we associate 0 with position x. By doing so, we are left with a new text array consisting only of zeros and ones. Finally, we look for occurrences of an $m \times m$ array containing only 1s. Therefore, the algorithm is essentially the same as in the one-dimensional case. But here, the relationship between positions is a bit more complicated. This is why relations $<_b$ and $<_t$ have been introduced. The following property of consistent positions is crucial for the correctness of the algorithm.

Consistency property (transitivity). Let $x <_t y <_t z$, or $x <_b y <_b z$. If x, y are consistent and y, z are consistent, then x, z are also consistent.

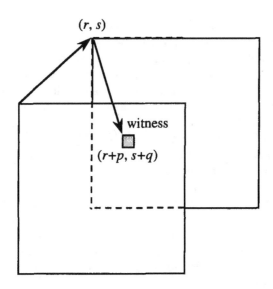

Figure 15.4: The second category of witnesses. The vector (r, s) is not a period, $(p, q) = WIT[r, s]$, and $PAT[p, q] \neq PAT[r + p, s + q]$.

According to relations $<_b$ and $<_t$, consistency refines to *bottom consistency* and *top consistency*. A set of positions is *bottom consistent* if for any two positions x, y of the set, such that $x <_b y$, the positions are consistent. Top consistent positions are defined similarly. It is clear that two elements are consistent iff they are top *and* bottom consistent. The same refers to sets of (pairwise) consistent positions.

Let R be a sub-rectangle of the text array T. The set S of positions in R is said to be *good* with respect to R if both positions in S are pairwise consistent, and there is no matching position within $R - S$. Let k be a column of the text array T. In the searching algorithm, we maintain the following invariant:

a good set of consistent positions in the columns $k, k + 1, \ldots, n'$ is known.

First, we construct good sets of consistent positions separately for each columns. This gives the invariant for $k = n'$. Then we satisfy the invariant for $k = n' - 1, n' - 2, \ldots, 1$. At completion we have a good set of consistent positions for the entire text array.

When processing the k-th columns, we run through consistent positions of this columns in a top-down fashion (see Figure 15.5). We maintain the following invariant:

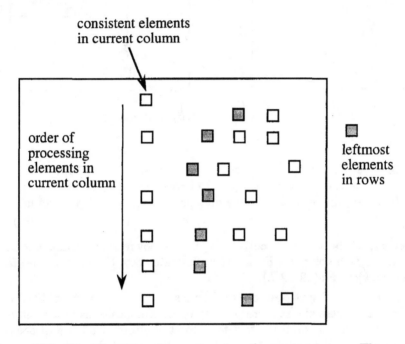

Figure 15.5: The situation when processing the next column. The current columns contains positions mutually consistent within this columns. Then, positions inconsistent with other columns are removed.

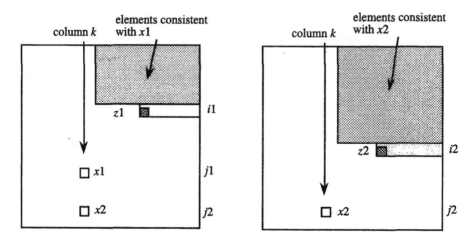

Figure 15.6: From $inv(x1, z1, k)$ to $inv(x2, z2, k)$.

$inv(x, z, k)$: z is the leftmost consistent position in its own row; let R be the rectangle composed of rows above z, and columns $k, \ldots, n'$; the set S of all remaining positions in R is a good set in R.

Let $x1$, $x2$ be two consecutive (in the top-down order) consistent positions in the k-th columns. Figure 15.6 illustrates how the process goes from $inv(x1, z1, k)$ to $inv(x2, z2, k)$.

The set of consistent positions in columns $k + 1, \ldots, n'$ is maintained as a set of stacks. These stacks correspond to rows. The positions in a given row, from left to right, are on their stack from top to bottom: the leftmost position in i-th row (in columns $k + 1, \ldots, n'$) is at the top of the i-th stack. The duels of $x1$ against elements of the i-th row are executed using the same stack procedure as in the string-matching algorithm by duels (see Chapter 13).

Assume that we start a given phase with position $x1$ in the k-th columns, and with position $z1$ in the $i1$ row (see Figure 15.6). The rows are processed top-down and left-to-right, starting with $z1$, and ending before or at the row containing $x1$. Initially $z = z1$. Then, assume that we consider a candidate z in a column to the right of $x1$. The basic operation is the duel between $x1$ and z. Three cases are possible:

1. both $x1$ and z survive (they are consistent); the *crucial point* is that we know at this moment (due to transitivity of consistency) that all candidates to the right of z and in the same row as z are consistent with

$x1$; we do not need to process them; we just go to the next row, starting with the leftmost candidate z (to the right of k-th column) in this row;

2. z is "killed" by $x1$ in the duel; we process the next candidate in the same row as z; if there is no such candidate we simply go on to the next row;

3. $x1$ is "killed" and z survives; then, the processing of $x1$ has been completed; $x1$ is removed as a candidate, $z1 = z$, and we start the next phase with the next candidate $x2$ below $x1$ in the same column as $x1$; if there is no such candidate, then the processing of the entire column has been completed; take $x1$ as the top most candidate in the $(k-1)$ column, and start processing column k-1.

Through this process, we obtain a set of consistent positions in the sense of the ordering $<_t$. If x, y are in this order, and are in the set, then they are consistent. After that, we again process the whole text array, but in a bottom-up manner, essentially performing the same algorithm as described above for the top-down ordering. The rows are again processed from left to right. The remaining set of positions is guaranteed to be bottom consistent. Thus, the final set S is a good consistent set.

The problem is reduced to "unary" pattern matching, in which the pattern consists of only one symbol, as follows. For each position x in the text array, find any position y in S such that x is in the range of y. Place the pattern at position y on the text array, and check if the symbol at x matches the corresponding symbol of the pattern. If "yes" associate "1" with position x. If "no", or if there is no such position y, associate "0" with x. In this way, we obtain a new array of zeros and ones. What remains is to search for a rectangular shape of size $m \times m$ containing only 1's inside the new array. This is straightforward, and is left to the reader. The above discussion gives a proof of the following statement.

Theorem 15.1 *If the witness table for the pattern array is computed, the search phase for the two-dimensional pattern matching can be done in linear time, independent of the size of the alphabet.*

The computation of the witness table given in the following employs a suffix tree. It takes a time that depends on the size of the alphabet, though it is linear with respect to the length of the pattern. This is due to the construction of suffix trees. In the computation of the witness table, the basic operation consists of checking the equality of two sub-rows of the pattern. This is executed on the suffix tree of the set of rows of the pattern, after it has been preprocessed for LCA queries.

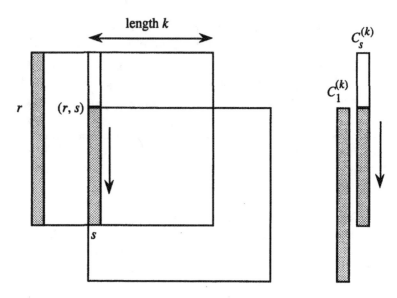

Figure 15.7: Computing a witness for position (r, s) of PAT.

Theorem 15.2 *The witness table for an $m \times m$ two-dimensional pattern on the alphabet A can be computed in $O(m^2 \log |A|)$ time.*

Proof. Let us examine the situation in which the witness for the position (r, s) of PAT is to be computed. Let $k = m - s + 1$. Assume that elements of the first and of the s-th columns are names of the rows of size k starting at the positions of columns to the right. Denote the resulting columns by $C_1^{(k)}$ and $C_s^{(k)}$ (see Figure 15.7).

Let ST be the suffix tree of all rows of the pattern. It takes $O(m^2 \log |A|)$ time to build this tree. Preprocess the tree in order to answer LCA queries (locating Lowest Common Ancestors) in constant time. Also consider the table $PREF$ as defined in Chapter 3.

We show that the table $PREF$ of the column $C_s^{(k)}$ with respect to $C_1^{(k)}$ can be computed in $O(m)$ time, once the tree ST is given. The computation of this table essentially reduces to the computation of the table of border lengths (see Chapter 3). We don't need actually the names of entries of the columns $C_s^{(k)}$ and $C_1^{(k)}$. These names represent sub-rows of length k. It is sufficient to make comparisons in constant time, hence, it is also sufficient to be able to quickly check the equality between two sub-rows of the same size k. This can be executed using LCA queries about the rows of the pattern array. First assume that (r, s) is a period of the pattern. Then $PREF[r] = m - r$. Otherwise, $PREF[r]$ gives the index of the row where the witness position is. To find the witness position it is sufficient to find the longest common prefix of two sub-rows of length k. This can be done using ST and LCA queries. ☐

15.2   Geometry of two-dimensional periodicities

This section presents some theoretical tools for the Galil-Park 2D-pattern-matching algorithm of Section 15.4. The proofs of several simple facts are omitted. Let PAT be a two-dimensional pattern of shape $m \times m$, with its rows and columns numbered by $1, 2, \ldots, m$. The vectors of PAT are denoted by π and β. We consider only two-dimensional vectors with integer components. Recall that a vector π is a *period* of PAT if $PAT[x] = PAT[x + \pi]$, whenever both sides are defined. If both sides are defined for at least one point x, then π is a nontrivial period. We also write that π is a 1D-period to emphasize its one-dimensional status. The main difference between 1-dimensional and 2-dimensional pattern matching lies in the different structures of periods of patterns. In two dimensions, some periods are inherently two-dimensional and are called 2D-periods.

A pair $\mu = (\pi, \beta)$ of non-collinear vectors is a 2D-period of PAT iff π and β are nontrivial periods, and each linear combination of π and β is a 1D-period of PAT. An equivalent formulation is: PAT can be extended to an infinite plane in which π and β are periods. By a linear combination we always mean a combination with integer coefficients, i.e., a vector $i.\pi + j.\beta$, in which i, j are integers. Two (or more) vectors are said to be collinear if they are in the same direction, which, in this case, does not necessarily mean that one is an integer combination of the other (others). Let us denote by $Lattice(\mu)$ the set of all linear combinations of π, β. The elements of $Lattice(\mu)$ are called the lattice points. Therefore, the pair μ is a 2D-period iff all elements of $Lattice(\mu)$ as vectors, are periods.

A vectors $\pi = (r, c)$ is said to be *small* iff its components r, c satisfy $|r| < d.m$ and $|c| < d.m$, in which $d = 1/16$. That is, a 2D-period is *small* if both its components are small vectors. The pattern PAT is called *periodic (lattice-periodic or 2D-periodic)* if it has a *small* 1D-period (2D-periods).

Remark. In one-dimensional string matching a linear combination of *small* 1D-periods is always a period. But this is not generally valid for two dimensions, even for non-negative combinations of collinear vectors (as well as for non-collinear vectors, of course). If all elements of the array PAT are the same letter except for a small number of elements closed to one fixed corner, then there are many 1D-periods, but there is no non-trivial 2D-period. The part around the corners is responsible for the irregularities.

The 2D-period $\mu = (\pi, \beta)$ is said to be *normal* if π is a quad-I period as in Figure 15.1 and β is a quad-II period as in Figure 15.4.

Lemma 15.1 [normalizing lemma] *If the pattern has a small 2D-period then it has a small normal 2D-period.*

A notion of divisibility for 2D-periods, $\mu 1, \mu 2$, is introduced as follows:

$$\mu 1 \mid \mu 2 \text{ iff } Lattice(\mu 1) \text{ includes } Lattice(\mu 2).$$

We also introduce the notion of a *smallest* 2D-period $\mu = period(PAT)$. It is a fixed small 2D-period of *PAT* that divides each other small 2D-period: in other words, $\mu \mid \mu'$ for each small 2D-period μ'. There are several ways to decide which μ is to be chosen, but any of them is good. Assume *PAT* is 2D-periodic. We define $period(PAT)$ as a small *normal* 2D-period (π, β), in which π is a quad-I small period of a minimal length (in case of ties the most horizontal vector is chosen), and β is a quad-II small period of a minimal length (in case of ties the most vertical vector is chosen). Lemma 15.1 guarantees that this definition makes sense. It can be proven that any small 1D-period corresponds to a point in $Lattice(PAT)$. This implies the following lemma.

Lemma 15.2 [2D-periodicity lemma] *(a) Assume $\mu 1, \mu 2$ are small 2D-periods of PAT. Then, there is a 2D-period μ such that $\mu \mid \mu 1$ and $\mu \mid \mu 2$.*
(b) Assume PAT is lattice-periodic and π is a small vector. Then π is a 1D-period of PAT
*   iff π is in $Lattice(period(PAT))$. Moreover, $period(PAT) \mid \mu$ for all small 2D-period μ.*

Observation. Assume we know which points of *PAT* are small sources (see below). Then, if it exists, $period(PAT)$ can be computed in $O(M)$ time independently of the alphabet.

Lemma 15.3 [overlap lemma] *Assume the patterns PAT_1 and PAT_2 are 2D-periodic subsquares of the same rectangle, that we have $period(PAT_1) = \mu 1$ and $period(PAT_2) = \mu 2$, in which $|\mu 1|, |\mu 2| < m/2 \leq m$. If PAT_1 and PAT_2 overlap on an $m \times m$ square, then $\mu 1 = \mu 2$.*

According to their periodicities, 2D-patterns are classified into four main categories:

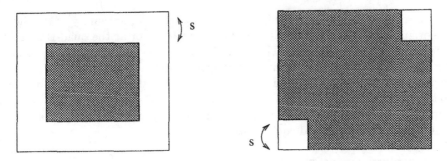

Figure 15.8: The shaded areas correspond to $Center_s(PAT)$ (left) and $Cut\text{-}Corners_s(PAT)$ (right).

- *non-periodic*: no small period at all,

- *lattice-periodic* (or *2D-periodic*): at least one small 2D-period,

- *radiant-periodic*: at least two non-collinear small 1D-periods, but not lattice-periodic,

- *line-periodic*: all periods in the same direction.

We have already defined quad-I periods and quad-II periods. We recall these definitions and introduce similar categories for so-called *sources*. The pattern is divided into four $m/2 \times m/2$ disjoint squares, called *quads*, and named quad I, quad II, quad III, and quad IV, according to the counter-clockwise ordering, and starting at the upper left corner. Therefore, quad I corresponds to the upper left square corner, and quad II corresponds to the lower left square corner.

For technical reasons it is convenient for these categories to be disjoint. So we assume that horizontal vectors are not quad II vectors, and vertical vectors are not quad-I vectors. As per this assumption, each 1D-period oriented from left to right is exactly of one type: a quad-I period or a quad-II period.

Let $Center_s(PAT) = PAT'$ be the central sub-array of PAT that results after "peeling off" the s boundary columns and rows (from top, down, right, and left). The shape of such a sub-array is $(m - 2s) \times (m - 2s)$.

We state the following lemma without proof.

Lemma 15.4 [radiant-periodicity lemma] *Assume that PAT is radiant-periodic. Then the array $Center_{2d.m}(PAT)$ is not radiant-periodic.*

A vector π can be identified as the point $\underline{\pi}$ of PAT, extremity of π, when its origin is at the quad-I top left corner (point$(0,0)$) or at the quad-II corner (point $(m,0)$) of the pattern. The point $\underline{\pi}$ is called the quad-I beginning point, or the quad-II beginning point, respectively, corresponding to π. If π is a period and the quad-I beginning point $\underline{\pi}$ is in PAT, then π is called a quad-I period, and $\underline{\pi}$ is called a quad-I *source*. Quad-II periods and sources are defined analogously. Using the terminology of sources, the periodicity type of pattern PAT can be characterized equivalently as follows:

- *non-periodic*: no small source;

- *lattice-periodic* (or *2D-periodic*): at least one quad-I source and one quad-II source;

- *radiant-periodic*: not lattice-periodic and at least two non-collinear small sources;

- *line-periodic*: all sources on the same line.

Let $\mu = (\pi, \beta)$ be a 2D-vectors. We say that two points x, y are μ-equivalent if $x - y$ is in $Lattice(\mu)$. A μ-path is a path consisting of edges that are vectors $\pi, -\pi, \beta$ or $-\beta$. The processing of certain difficult patterns is executed by exploring some simple geometry of paths on a lattice generated by two vectors π, β belonging to the same quadrant. If we have two points x, y containing distinct symbols and we have a (π, β)-path from x to y within the pattern, then one of the edges of the path gives a witness of non-periodicity to one of the vectors π or β. This is due to the fact that the initial and terminating positions do not match, so there should be a mismatch "on the way" from x to y. The length of the path is the number of edges it contains.

Observation. The basic difficulty with such an approach is the length of the path. It may be that μ is a small 2D-period, but the length of a shortest μ-path between two μ-connected points of an $m \times m$ square is quadratic. Consider, for example, $\mu = ((m/2, 1), (1, 0))$, $x = (m/2, 0)$ and $y = (m/2, m - 1)$.

Despite the previous observation, we can find useful short paths in some situations, as shown in the next lemma. Let $Cut\text{-}Corners_s(R)$ be the part of array R without top-right and bottom-left corner squares of shape $s \times s$, see Figure 15.8.

Lemma 15.5 [linear-path lemma] *Assume π, β are quad-I vectors of size at most k. Let S be a subsquare of size $k \times k$ of a large square R, and let x be the point that is the bottom-left corner or top-right corner of S. Assume x is inside $Cut\text{-}Corners_{2k}(R)$. Then, there is a linear-length μ-path inside R from*

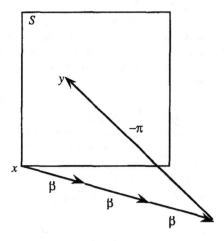

Figure 15.9: Illustration of Lemma 15.5: a *short* path from x to $y \in S$.

x to a point $y \neq x$ in S, see Fig 15.9. Such a path can be computed in $O(k)$ time.

Proof. We can assume, without loss of generality, that x is the quad-II corner of S, and that π and β are quad-I vectors of size at most k. Moreover, we can assume that the array R is of shape $3k \times 3k$, and S is the square of size $k \times k$ at the top-left corner of R. Then, x is the point of R at position $(k, 0)$. Assume that β is a more horizontal vector than π. We find a path and a point y by the algorithm *GREEDY* presented below.

Algorithm *GREEDY*
  $y := x$;
  repeat
    if $y - \pi$ is outside the area R **then** $y := y + \beta$
    else $y := y - \pi$;
  until y is in S;

We prove that the algorithm *GREEDY* terminates successfully after a linear number of iterations and generates the required path. Consider the lines $L_0, L_1, L_2, \ldots$, in which L_h is the line parallel to π and that contains the points $x + h\beta$. Then, points $x + i\beta + j\pi$ (i, j integers) belong to the line L_i. If some line L_i cuts the two horizontal borders of S, or its two vertical borders, then the segment of the line that is inside S is longer than π. Thus, $x + i\beta + j\pi$ belongs to S for some (negative) integer j. If each line L_i cuts both a horizontal and vertical border of S then let i be such that lines L_i and L_{i+1} surround the diagonal segment of S; it can then be proven that, either there is a point $x + i\beta + j\pi$ in S or a point $x + (i+1)\beta + j'\pi$ in S. Values of variable y in the algorithm are points of a μ-path inside R because if "$y - \pi$ not in R," $y + \beta$ is in R. We have yet to explain why the path has linear length, which at the same time proves that the algorithm works in $O(k)$ time. Let $\pi = (r, c)$ and $\beta = (r', c')$. The point $x + r\beta - r'\pi$ is on the same column as x, and can be the point y if it is in S. It is clear that the μ-path followed by the algorithm is entirely (except maybe its last edge) inside the triangle $(x, x+r\beta, x+r\beta-r'\pi)$. Thus, the length of the μ-path followed by the algorithm is no longer than $r+r'$ which is $O(k)$. □

We introduce a special type of duels called here *long-duels*. Assume we have small quad-I vectors π, β and a point x at distance at least $2k$ from quad-II and quad-IV corners. Assume also that if y is any point such that $y - x$ is a small quad-II vectors, then $PAT[x] \neq PAT[y]$. The procedure *long-duels*(π, β, x) "kills" one of the vectors π, β and finds its witness. It works as follows: a (π, β)-path from a given point x to some point y in the pattern is found by the algorithm of the linear-path lemma. The path consists of a linear number of edges. End points x and y contain distinct symbols. Therefore, one of the edges on the path consists of a linear number of edges. End points x and y contain distinct symbols. Therefore, one of the edges on the path gives a witness for π or β. In doing so, one of the potential small periods π or β is "eliminated" in linear time.

Theorem 15.3 [long-duel theorem] *Assume we have a set X of small quad-I vectors, in which $|X| = O(m)$, and that we are given a position x inside Cut-Corners$_{2d.m}(PAT)$. Assume also that if y is any point such that $y - x$ is a small quad-II vector, then $PAT[x] \neq PAT[y]$. Then, in linear time, by using long duels, we can find witnesses for all small quad-I vectors (except maybe for a set of vectors on a same line L).*

Proof. We run the following instruction.

> **while** X is non-empty **do begin**
>   take any element β from X; add β to Y, and delete β from X;
>   **while** there are two non-collinear vectors π, β in Y **do**
>   execute *long-duel*(π, β, x) and delete the "loser" from Y;
> **end;**

We keep the elements of X that have not been eliminated so far in the set Y. Initially Y is empty. The invariant of the loop is: non-null witnesses for all elements not in the current sets X or Y are computed, and all elements of Y are on the same line L. Altogether, the execution time is $O(m^2)$ and alphabet independent. At the end, all remaining vectors (with non-null witnesses up to now) are on a same line L. This completes the proof. □

Let us call the algorithm of the long-duel theorem the *long-duel algorithm*. There is a natural analogue of the theorem for quad-II small vectors, and for quad–I and quad-III corners. The crucial point is played by the following suffix-testing problem: given m strings $x_1, \ldots, x_m$ of total size $O(m^2)$, compute the $m \times m$ table *Suf-Test* defined as follows:

Suf-Test$[i, j]$ = nil if the i-th string is a suffix of the j-th string,
Suf-Test$[i, j]$ = positions of the rightmost mismatch otherwise.

The algorithm is given below as Algorithm *Suffix-Testing*. We sketch its rough structure to show that it runs in linear time independently of the alphabet. It is sufficient to compute for each pair (i, j) the length $SUF[i, j]$ of the longest common suffix of x_i and x_j. The algorithm can be easily implemented to work in $O(m^2)$ time. The main point in the evaluation of the time complexity is that if a position participates in a positive comparison (when two symbols match), then this position is never inspected again. When we process a given word x_k and compute $SUF[k, j]$ for $j > k$, then we first look at $SUF[i, j]$, in which $i = MAX[j, k - 1]$, $1 \le i \le k$, is the index that maximizes $SUF[i, j]$, and then at $SUF[i, k]$. These data are available at this moment, due to invariant. The word x_j is scanned backward starting from position $SUF[i, j]$. The pointers only go backward. This proves the following.

Theorem 15.4 [suffix-testing theorem] *The suffix-testing problem related to m strings of total size $O(m^2)$ can be solved in $O(m^2)$ time, independently of the alphabet.*

Algorithm *Suffix-Testing*;
    assume strings $x_1, \ldots, x_m$ in increasing order of their lengths;
    { *invariant(k)*: for all i, j, $1 \le i \le k$, $1 \le j \le m$,
        $SUF[i, j]$ is computed, and, for each j, $1 \le j \le m$, we know
        $i = MAX[j, k]$ the index $i \le k$ that maximizes $SUF[i, j]$ }
    make *invariant(1)* ;
    for $k := 2$ **to** m **do**
        make *invariant(k)* using *invariant(k − 1)* ;

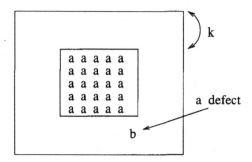

Figure 15.10: A *mono-central* pattern.

15.3 ∗ Patterns with large monochromatic centers

The alphabet-independent linear-time computation of 2D-witness tables is quite technical, hence the present and following sections may be considered as optional. In this section, we present an alphabet-independent linear-time computation of witnesses for special patterns, the "large" central part of which is "monochromatic." The pattern PAT is called *mono-central* if all symbols lying in $Center_k$ are equal to the same letter a of the alphabet, for $k < 3m/8$. Then, the central sub-array of PAT of size at least $m/4 \times m/4$ is *monochromatic*. A position containing a letter different from letter a is called a *defect*, see Figure 15.10

Assume the existence of at least one defect (otherwise the preprocessing is trivial). Opposite corners of PAT are the corners lying on the same forward or backward diagonal of PAT (quad-I and quad-III, or quad-II and quad-IV corners). A *mono-central* pattern PAT is called a *corner* if there is a pair of opposite corners x, y of PAT such that each defect can be reached by at most two small vectors from x or y. The *corner* patterns are the most difficult with respect to their witness computation, because they can be radiant-periodic. The non-corner patterns are simpler to deal with, due to the following observation.

Observation. If PAT is a periodic *non-corner mono-central* pattern, then, PAT is non-periodic or line-periodic (therefore, PAT is not radiant-periodic).

Lemma 15.6 [subsquare lemma] *Let us assume that PAT is mono-central, and that there is a defect inside the area $R = Cut\text{-}Corners_{2k}(PAT)$. Then there is a defect position x within R satisfying one of the following conditions:*

(1) x is a quad-I or is a quad-III corner of a $k \times k$ subsquare S containing no defect position strictly within S;
(2) x is a quad-II or is a quad-IV corner of a $k \times k$ subsquare S containing no defect position at all, except x.

Proof. Take a defect point in R closest to the center of PAT.   □

Theorem 15.5 [non-corner theorem] *The witness table for all small vectors of mono-central non-corner pattern PAT can be computed in $O(m^2)$ time.*

Proof. Consider the defect z closest to the center of PAT. Assume without loss of generality that z is in quadrant II. The position z is the witness (of non-periodicity) for all small quad-II vectors, except perhaps for vertical vectors. The case of vertical and horizontal periodicities is very easy to process, therefore, we assume that all witnesses for vertical and horizontal vectors are computed and PAT is not vertically nor horizontally periodic. The set of potential small quad-I periods is sparsified using vertical duels in columns. Afterward, in quadrant I we have only a linear number of candidates for small periods. Denote by X the set of these candidates. To compute witnesses for quad-I small vectors it is sufficient to find a point x implied by the subsquare lemma. If condition (1) of this lemma holds, then x itself is the witness for all quad-I vectors that are not vertical nor horizontal vectors. Otherwise, x is "good" to apply the *long-duel* theorem. The case of a horizontally- or vertically-periodic pattern can be easily processed. This completes the proof. □

Theorem 15.6 [corner theorem] *Consider an $m \times m$ array PAT that is a mono-central corner pattern. Then, witnesses for all vectors of size at most $m/8$ can be computed in $O(m^2)$ time.*

Proof. Assume that opposite corners from the definition of corner patterns are quad-II and quad-IV corners. Then, all defects are closed to quad-II and quad-IV corners. These corners are separated by a large area of non-defects. Therefore, we can compute periods and witnesses separately with respect to each corner. Hence, without loss of generally, we can assume that all defect are close to the quad-II corner, and, in particular, that there is no defect that contains the same symbol a in quadrants I, III, and IV. Assume PAT contains at least one defect in quadrant II. PAT obviously has no small quad-II periods, since the rightmost defect gives witnesses against all quad-II vectors. We show how to compute witnesses for small quad-I vectors.

Let $PAT1$ be the following transformation of the pattern. In each row replace all symbols by a, except the rightmost non-a symbol of each row. Replace these rightmost non-a symbols by a special symbol $. Let X be the set of positions containing the symbol $; call them *special* positions.

For x in X denote by $string(x)$ the word in PAT consisting of the part of the row containing x from the left side up to x (including x). Let π be a vector of size at most $m/4$. Then, it is easy to see the following:

Claim. π is not a period iff (1) π is not a period in $PAT1$, or (2) for two positions w, y in X we have: $y - x = \pi$ and $string(x)$ is not a suffix of $string(y)$ (then, a witness for π is given by a mismatch between $string(x)$ and $string(y)$).

The computation of all witness for vectors of size at most k in $PAT1$ is rather simple. Only quad-I vectors are to be processed. Assume there is no small vertical period. Then, the set of potential small quad-I periods is sparsified using duels in columns. Afterward a linear number of candidates remains. Each of them is checked against all (linear number) symbols at special positions in a naive way. The witnesses arising from condition (2) are computed directly using the *Suffix-Testing* algorithm. This completes the proof. $\square$

We extend the definition of *defects*. Assume a mono-central pattern PAT has a lattice-periodic central sub-array C of size at least $m/4$. We say that x is a *lattice-defect* if x does not agree with (contains a symbol different from) any point y in C that is lattice-equivalent to x. Let $Mono(PAT)$ be the pattern in which all positions that are not *lattice-defects* are replaced by the same special symbol. We omit the proof of the following simple lemma.

Lemma 15.7 [mono lemma] *If a small vector π is in the lattice generated by the smallest period of C, then π is a period of PAT iff π is a period of $Mono(PAT)$.*

15.4 ∗ A version of the Galil-Park algorithm

Recall that the periodicity type of a sub-array depends on its size. When we say that the witnesses for a given array or sub-array are computed we mean the witnesses, if any, for all vectors that are small according to the size of the presently considered array.

Lemma 15.8 [line lemma] *Assume we have a set S of points, in a fixed quadrant of PAT, such that they are all on the same line L. Then, we can check which of them correspond to periods, and compute witnesses, wherever they are, in $O(m^2)$ time independently of the alphabet.*

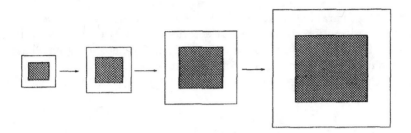

Figure 15.11: The graphical illustration of the performance of Galil-Park al-gorithm, the witnesses are computed for small vectors in a central sub-pattern, then they are iteratively computed (using procedures *Extend*) to geometrically larger sub-arrays (resulting from the *Large-Extend* operation).

Proof. The proof reduces to the computation of witness tables for m one dimensional strings of size $2m$ each. Let L_i be all lines parallel to L; take m pairs of lines (L_i, K_i), in which K_i is parallel to L_i and the distance between K_i and L_i equals the distance between the point $(0,0)$ and L. Each line is taken as a string of symbols. For each i, lines L_i and K_i are concatenated, and witnesses for these strings are computed using a one-dimensional classical algorithm.   □

Assume C is a central sub-array of shape $s \times s$, in which $s \leq m$. Denote by *Large-Extend*$(C) = D$, in which D is a twice larger central sub-array of shape $2s \times 2s$. If ever $2s > m$, we define *Large-Extend*$(C) = PAT$. Observe that a small period with respect to *Large-Extend*(C), in case *Large-Extend*$(C) \neq PAT$, means a vector of size at most $2d.s$ (recall that $d = 1/16$). *Large-extend*(C) is twice as large as C except maybe at the last iteration of the algorithm. The reason for such irregularity is that while making duels for small vectors in D we use mismatches in C, and we should guarantee that D is large enough with respect to C, and that vectors (taking part in a duel) starting in C do not go outside the pattern PAT. Also define *Small-Extend*(C) as a central sub-array of shape $3s/2 \times 3s/2$. Due to Lemma 15.4, if *Large-Extend*$(C) \neq PAT$ and if it is radiant-periodic, then *Small-Extend*(C) is not radiant-periodic. This saves one case (radiant-periodic) in the algorithm: C is never radiant-periodic.

Before each iteration in GP algorithm witnesses and periods are already known for a central sub-array C, which shape is $s \times s$. The witnesses for a larger central sub-array D are computed, in which $D =$ *Large-Extend*(C), or $D =$ *Small-Extend*(C) in the case *Large-extend*(C) is radiant-periodic and *Large-Extend*$(C) \neq PAT$. In the latter case, D has shape $3s/2 \times 3s/2$, and

Lemma 15.4 guarantees that D is not radiant-periodic. Then C is set to D and the next iteration starts. We first describe three procedures that compute witnesses for small vectors in $D = Large\text{-}Extend(C)$, depending on what type of periodicity is in C. In Figure 15.11 the sub-array C is shaded. At the next stage of Galil-Park algorithm we will have $(C, D) := (D, \, Large\text{-}Extend(D))$.

---

procedure *Nonperiodic-Extend(C)*;
    the witness table of C is used to make duels between
        candidates for small periods in D;
    after dueling, only a constant number of candidates remains,
        their witnesses are computed in a naive way;

---

procedure *Lattice-Periodic-Extend(C)*;
    consider the areas $Q1, Q2$ of candidates of small
    (with respect to D) periods in, respectively, quad I and
    quad II of D; divide each area into $d.s \times d.s$ subsquares;
    in each smaller subsquare **do begin**
        make duels between candidates using witnesses from C;
        { only candidates on the same line survive }
        apply the algorithm from the *Line Lemma*;
    end

Three disjoint cases are considered in the Galil-Park algorithm depending on whether C is non-periodic, lattice-periodic, or line-periodic. The first case (non-periodic) is very simple.

At each iteration we spend $O(r^2)$ time, in which r is the size of the actual array D; this size grows at least by a factor $3/2$ at each iteration. Altogether, the time is linear with respect to the total size of the pattern, as the sum of a geometric progression.

---

procedure *Line-Periodic-Extend*(C);
  { C is lattice-periodic }
    let $\mu = period(C)$;
    for each small candidate period $\pi \in Lattice(\mu)$ **do**
        find a μ-equivalent point y in quad I or quad II of C,
          then the witness corresponding to y gives a witness for x;
    for each small candidate period $\pi \in Lattice(\mu)$ **do**
        compute witness of π in $Mono(PAT)$; { Mono Lemma }
        { use algorithms from *Non-corner* or *Corner Theorem* }

When the witness table is eventually computed, the Amir-Benson-Farach searching phase can be applied. Altogether we have proven the following result.

Theorem 15.7 *There is a 2D-pattern-matching algorithm which time complexity is linear in the size of the input, and independent of the alphabet (including the preprocessing).*

Algorithm GP; { modified Galil-Park algorithm; }
    { computes witnesses for all small vectors of PAT }
  C :=an initial constant-size non radiant-periodic
    central sub-array of PAT;
  compute the witness table of C in $O(1)$ time;
  $D := Large\text{-}Extend(C)$;
  while $C \neq PAT$ **do begin** { main iteration }
    if C is non-periodic **then** *Nonperiodic-Extend*(C)
    else if C is line-periodic **then** *Line-Periodic-Extend*(C)
    else *Lattice-Periodic-Extend*(C);
    if $D \neq PAT$ and D is radiant-periodic **then**
        $D := Small\text{-}Extend(C)$
    else begin $C := D$; $D := Large\text{-}Extend(C)$ **end**;
  end { of main iteration }

Bibliographic notes

The linear-time searching algorithm of Section 15.1 is from Amir, Benson, and Farach [ABF 92a]. It is quite surprising that this is the first alphabet-independent linear-time algorithm, because, in the case of strings, the first algorithm satisfying the same requirements is the algorithm of Morris and Pratt [MP 70]. The gap between these is more than twenty years! Furthermore, the preprocessing phase of the algorithm in [ABF 92a] is not alphabet-independent. The rest of the chapter is an adaptation of the results of Galil and Park in [GP 92b], it is a version of Galil-Park algorithm presented in [CR95].

The question of periodicities for two-dimensional patterns is discussed in several papers, particularly in [AB 92], [GP 92], and [GP 93]. A constant-space 2-dimensional pattern-searching algorithm has been designed in [CGPR95], using properties of periodicities.

Chapter 16

Parallel text algorithms

We present several *polylogarithmic*-time parallel algorithms using a high level description of the *Parallel Random Access Machine* (PRAM). A lot of research has been done on the so-called *optimal parallel algorithms*, the ones that achieve linear *total work* (the product of the number of processors by the parallel time). Optimal parallel text algorithms are Vishkin's algorithm and algorithms using the splitting technique (also known as *pseudoperiod technique*). In practice the polylogarithmic factors for the total work are not so important, especially if the total work is $O(n^{1+\alpha})$ for $\alpha > 0$. Example of such algorithms are parallel construction of Huffman trees and computation of edit distance. For these problems the total work of known polylogarithmic-time algorithms is far from linear. But the reduction below cubic work provides beautiful algorithms.

16.1   The abstract model of parallel computing

Concerning parallel computations, a very general model is assumed, since we are interested mainly in exposing the parallel nature of some problems without going into the details of the parallel hardware. The parallel random access machine (PRAM), a parallel version of the random access machine, is used as a standard model for presentation of parallel algorithms.

The PRAM consists of a number of processors working synchronously and communicating through a common random access memory. Each processor is a random access machine with the usual operations. The processors are indexed by consecutive natural numbers, and synchronously execute the same central program; but, the action of a given processor also depends on its number (known to the processor). In one step, a processor can access one memory

location. The models differ with respect to simultaneous access to the same
memory location by more than one processor. For the CREW (concurrent
read, exclusive write) variety of PRAM machine, any number of processors
can read from the same memory location simultaneously, but write conflicts are
not allowed: no two processors can attempt to write simultaneously into the
same location. CRCW (concurrent read, concurrent write) denotes the PRAM
model in which, in addition to concurrent read, write conflicts are allowed:
many processors can attempt to write into the same location simultaneously
but only if they all attempt to write the same value.

There is no generally accepted universal language for the presentation of
parallel algorithms. The PRAM is a rather idealized model. We have chosen
this model as the best one suitable for the presentation of algorithms, and
especially for the presentation of the inherent parallelism of some problems.
It would be difficult to adequately present these algorithms with languages
oriented toward concrete existing hardware of parallel computers. Moreover,
the PRAM model is widely accepted in the literature on parallel computation
on texts.

Parallelism will be expressed by the following type of parallel statement:

for all $i \in X$ **do in parallel** action(i).

The execution of this statement consists in

- assigning a processor to each element of X,

- executing in parallel by assigned processors the operations specified by
  action(i).

Usually the part "$x \in X$" looks like "$1 \leq i \leq n$" if X is an interval of integers.

Methodologically one can apply two different approaches for constructing
efficient parallel algorithms:

(1) translation into a parallel version of a known sequential algorithm,

(2) design of a new algorithm with a good parallel structure.

Method (1) works well in the case of almost optimal parallel string-matching
algorithms, and square finding. The known *KMR* algorithm, and the Main-
Lorentz algorithm (for squares) already have a good parallel algorithmic structure. However method (1) works poorly in the case of edit distance and Huffman coding, for example.

The PRAM model is best suited to work with tree-structured objects or
tree-like (recursive) structured computations. As an introduction we show such

a type of computation on one of the basic parallel operations known as *prefix computation.*

Given a vector x of n values the problem is to compute all prefix products:

$$y[1] = x[1], \; y[2] = x[1] \otimes x[2], \; y[3] = x[1] \otimes x[2] \otimes x[3], \; \ldots$$

Let us denote by *prefprod*(x) the function that returns the vector y as value. We assume that $\otimes$ is an associative operation computable on a RAM machine in $O(1)$ time. We also assume for simplicity that n is a power of two. The typical instances of $\otimes$ are arithmetic operations $+$, min and max. The parallel implementation of *prefprod* works as follows.

Lemma 16.1 *Parallel prefix computation can be accomplished in $O(\log n)$ time with $n/\log n$ processors.*

Proof. The algorithm above computes *prefprod*(x) in $O(\log n)$ time and uses n processors. The reduction of the number of processors by a factor $\log n$ is technical. We partition the vector x into segments of length $\log n$. A processor is assigned to each segment. All these processors simultaneously compute all prefix computations locally for their segments. Each processor does so using a sequential process. We then compress the vector by taking a representative (say the first element) from each segment. A vector x' of size $n/\log n$ is obtained. The function *prefprod* is applied to x' (now $n/\log n$ processors suffice because of the size of x'). Finally, all processors assigned to segments update values for all entries of their own segments using a (globally) correct value of the segment representative. This takes again $O(\log n)$ time, but uses only $n/\log n$ processors. □

```
function prefprod(x); { the size of x is a power of two }
    n := size(x);
    if n = 1 then return x else begin
        x₁ := first half of x; x₂ := second half of x;
        for each i ∈ {1, 2} do in parallel
            yᵢ := prefprod(xᵢ);
        midval := y₁[n/2];
        for each j, 1 ≤ j ≤ n/2, do in parallel
            y₂[j] := midval ⊗ y₂[j];
        return concatenation of vectors y₁ and y₂;
    end
```

16.2   Parallel string-matching algorithms

Suppose that v is the shortest prefix of the pattern that is a period of the pattern. If the pattern is periodic (vv is a prefix of the pattern) then vv^- is called the *non-periodic part of the pattern* (v^- denotes the word v with the last symbol removed). We omit the proof of the following lemma, which justifies the name "non-periodic part" of the pattern.

Lemma 16.2 *If the pattern is periodic (it is twice as long as its period) then its non-periodic part is non-periodic.*

The witness table (see Chapter 13) is relevant only for the non-periodic pattern. So, it is easier to deal with non-periodic patterns. We prove that such assumption can be done without loss of generality, by ruling out the case of periodic patterns.

Lemma 16.3 *Assume that the pattern is periodic, and that all occurrences (in the text) of its non-periodic part are known. Then, we can find all occurrences of the whole pattern in the text*
(i) in $O(1)$ time with n processors in the CRCW PRAM model, (ii) in $O(\log m)$ time with $n/\log m$ processors in the CREW PRAM model.

Proof. We reduce the general problem to unary string matching. Let $w = vv^-$ be the non-periodic part of the pattern. Assume that w starts at position i on the text. By a segment containing position i we mean the largest segment of the text containing position i and having a period of size $|v|$. We assign a processor to each position. All these processors simultaneously write 1 into their positions if the symbol at distance $|v|$ to the left contains the same symbol. The last position containing 1 to the right of i (all positions between them also contain ones) is the end of the segment containing i. Similarly, we can compute the first position of the segment containing i. It is easy to compute it optimally for all positions i in $O(\log m)$ time by a parallel prefix computation. The constant-time computation on a CRCW PRAM is more advanced; we refer the reader to [BG 90]. Some tricks are used by applying the power of concurrent writes.   □

Now we can assume that the pattern is non-periodic. We consider the witness table used in Chapter 13 for two sequential string-matching algorithms: by duels and by sampling. The parallel counterparts of these algorithms are presented. We skip the complicated proof of the preprocessing part of Vishkin algorithm.

Lemma 16.4 *The witness table can be computed in $O(\log^2 n)$ time on a CREW PRAM using $O(n/\log^2 n)$ processors.*

Recall that a position l on the text is said to be *in the range of* a position k if $k < l < k + m$. We say that two positions $k < l$ on the text are *consistent* if l is not in the range of k, or if $WIT[l - k] = 0$. If the positions are not consistent, then, in constant time we can remove one of them as a candidate for a starting position of the pattern using the operation *duel* (see Chapter 13).

Let us partition the input text into *windows* of size $m/2$. Then, the duel between two positions in the same window eliminates at least one of them. The position that "survives" is the value of the duel. Define the operation $\otimes$ by

$$i \otimes j = duel(i, j).$$

The operation $\otimes$ is "practically" associative. This means that the value of $i_1 \otimes i_2 \otimes i_3 \otimes \ldots \otimes i_{m/2}$ depends on the order of multiplications, but all values (for all possible orders) are good for our purpose. We need any of the possible values.

Once the witness table is computed, the string-matching problem reduces to instances of the parallel prefix computation problem. We have the following algorithm.

Algorithm *Vishkin-string-matching-by-duels*;
    consider windows of size $m/2$ on *text*;
    { sieve phase }
    for each window **do in parallel**
        { $\otimes$ can be treated as if it were associative }
        compute the surviving position $i_1 \otimes i_2 \otimes i_3 \otimes \cdots \otimes i_{m/2}$
        where $i_1, i_2, i_3, \ldots, i_{m/2}$ are consecutive positions
        in the window;
    { naive phase }
    for each surviving position i **do in parallel**
        check naively an occurrence of *pat* at position i
        using m processors;

Theorem 16.1 *Assume we know the witness table and the period of the pattern. Then, the string-matching problem can be solved optimally in $O(\log m)$ time with $O(n/\log m)$ processors of a CREW PRAM.*

Proof. Let i_1, i_2, i_3, ... , $i_{m/2}$ be the sequence of positions in a given window. We can compute $i_1 \otimes i_2 \otimes i_3 \otimes \cdots \otimes i_{m/2}$ using an optimal parallel algorithm for the parallel prefix computation. Then, in a given window, only one position survives; this position is the value of $i_1 \otimes i_2 \otimes i_3 \otimes \cdots \otimes i_{m/2}$. This operation can be executed simultaneously for all windows of size $m/2$. For all windows, this takes $O(\log m)$ time with $O(n/\log m)$ processors of a CREW PRAM. Afterward, we have $O(n/m)$ surviving positions altogether. For each of them we can check the match using $m/\log m$ processors. Again, a parallel prefix computation is used to collect the result, that is, to compute the conjunction of m Boolean values (match or mismatch, for a given position). This takes again $O(\log m)$ time with $O(n/\log m)$ processors. Finally, we collect the $O(n/m)$ Boolean values using a similar process.   □

Corollary 16.1 *There is an $O(\log^2 n)$ time parallel algorithm that solves the string-matching problem (including preprocessing) with $O(n/\log^2 n)$ processors of a CREW PRAM.*

Algorithm *Vishkin-string-matching-by-sampling*;
    consider windows of size $m/2$ on *text*;
    { sieve phase }
    for each window **do in parallel begin**
        for each position i in the window **do in parallel**
            kill i if the sample does not occur at i;
        kill all surviving positions in the window,
        except the first and the last;
        eliminate one of them in the field of fire of the other;
    end;
    { naive phase }
    for each surviving position i **do in parallel**
        check naively an occurrence of *pat* starting at i
        using m processors;

The idea of deterministic sampling was originally developed for parallel string matching. In Chapter 13, a sequential use of sampling is shown. Let us recall the definition of the good sample. A sample S is a set of positions on the pattern. A sample S occurs at position i in the text if $pat[j] = text[i+j]$ for each j in S. A sample S is called a *deterministic sample* if it is small ($|S| = O(\log m)$), and if it has a large field of fire (a segment $[i-k..i+m/2-k]$). If the sample occurs at i in the text, then positions in the field of fire of i, except i, cannot be matching positions. The important property of samples is that if we have two positions of occurrences of the sample in a window of size $m/2$. Then one "kills" the other: only one can possibly be a matching position.

In the sieve phase of the algorithm, we use $n \log m$ processors to check a sample match at each position. In the naive phase, we have $O(n/m)$ windows, and, in each window, m processors are used. This proves the following.

Theorem 16.2 *Assume we know the deterministic sample and the period of the pattern. Then, the string-matching problem can be solved in $O(1)$ time with $O(n \log m)$ processors in the CRCW PRAM model.*

The deterministic sample can be computed in $O(\log^2 m)$ time with n processors using a direct parallel implementation of the sequential construction of deterministic samples.

16.3 ∗ Splitting technique

Several open problems have been cracked using the approach of *splitting* a string into disjoint subsequences. The technique was originally known as the *pseudoperiod technique* for certain reasons. A small sample string z is selected and a binary string $occur(z, pat)$ representing all occurrences of z in pat is produced (string occurrences are ones, other positions are zeros). The smallest period of $occur(z, pat)$ is a *pseudoperiod* of pat. As a side effect, witnesses are computed for all relevant positions which are not multiples of the pseudoperiod. The elegancy of the method is obscured by many technical details. We only sketch some main ideas. The advantage is the (small) reduction of parallel time. We list the four most interesting problems that have been solved positively with this approach:

1. existence of a deterministic optimal string-matching algorithm working in $O(\log m)$ time on a CREW PRAM,

2. existence of a randomized string-matching algorithm working in constant time with a linear number of processors on a CRCW PRAM,

3. existence of an $O(\log n)$-time string-matching algorithm working on a *hypercube* computer with a linear number of processors,

4. existence of an $O(n^{1/2})$-time string-matching algorithm running on a *mesh-connected* array of processors.

All above problems can be solved using the pseudoperiod technique.

Theorem 16.3 *There exist efficient parallel algorithms for each of the four problems listed above.*

The technique also provides a new optimal string-matching algorithm working in $O(\log \log n)$ time on a CRCW PRAM. Recall that the CRCW PRAM is the weakest model of the PRAM with concurrent writes (whenever such writes occur the same value is written by each processor), and the CREW PRAM is the PRAM without concurrent writes. Vishkin algorithm works in $O(\log^2 n)$ time if implemented on a CREW PRAM. An optimal $O(\log n \log \log n)$ time algorithm for the CREW PRAM model was presented earlier by Breslauer and Galil. The full proof of Theorem 16.3 is beyond the scope of the book, we only point the most crucial features in the algorithms. The power of the splitting technique is related to the following recurrence relations:

$(*)$ $time(n) = O(\log n) + time(n^{1/2})$,

$(**)$ $time(n) = O(1) + time(n^{1/2})$.

Claim 1. The solutions of recurrence relations $(*)$ and $(**)$ satisfy, respectively:

$$time(n) = O(\log n) \text{ and } time(n) = O(\log \log n).$$

Let P be a pattern of length m. Its witness table WIT is computed only for the positions inside the interval $FirstHalf = [1..m/2]$. We say that a set S of positions is k-regularly *sparse* if S is the set of positions i inside $FirstHalf$ such that $i \bmod k = 1$. If S is regularly sparse then let $sparsity(S)$ be the minimal integer k for which S is k-regularly sparse. For $1 \leq q \leq k$ let us note:

$$P^{(q)} = P(q)P(k+q)P(2k+q)P(3k+q)\ldots$$

and

$$SPLIT(P,k) = \{P^{(q)} \ : \ 1 \leq q \leq k\}.$$

Example.

$$SPLIT(a\,b\,a\,c\,b\,d\,a\,b\,a\,d\,a\,a,\ 3) = \{a\,c\,a\,d,\ b\,b\,b\,a,\ a\,d\,a\,a\}.$$

Assume S is a k-regularly sparse set of positions. Denote by $COLLECT(P,k)$ the procedure that computes values of the witness table for all positions in S, assuming that the witness tables for all strings in $SPLIT(P,k)$ are known.

Claim 2. Assume the witness tables for all strings in $SPLIT(P,k)$ are known. Then, $COLLECT(P,k)$ can be implemented by an optimal parallel algorithm running in $O(\log m)$ time on a CREW PRAM, and in $O(1)$ on a CRCW PRAM.

The next fact is more technical. Denote by $SPARSIFY(P)$ the function that computes the witness table at all positions in *FirstHalf* except at a set S that is k-regularly sparse. The value returned by the function is the sparsity of S; when $k > m/2$, S is empty. In fact, the main role of the function is the sparsification of non-computed entries of the witness table.

Claim 3. $SPARSIFY(P)$ can be computed by an optimal parallel algorithm in $O(\log m)$ time on a CREW PRAM, and in $O(1)$ on a CRCW PRAM. The value k of $SPARSIFY(P)$ satisfies $k \geq m^{1/2}$.

The basic component of the function $SPARSIFY$ is the function $FINDSUB(P)$ that finds a non-periodic subword z of P of size $m^{1/2}$, or reports that there is no such subword. It is easy to check whether the prefix z' of size $m^{1/2}$ is non-periodic or not (we have a quadratic number of processors with respect to $m^{1/2}$); if z' is periodic we find the continuation of the periodicity and take the last subword of size $m^{1/2}$. The computed segment z can be preprocessed (its witness table is computed). Then, all occurrences of z are found, and, based on these, the sparsification is performed.

We only present how to compute witness tables in $O(\log m)$ time using $O(m \log m)$ processors. The number of processors can be further reduced by a logarithmic factor, which makes the algorithm optimal. The algorithm is illustrated by the following procedure *Compute-by-Splitting*.

```
procedure Compute-by-Splitting(P);
    k := 1;
    while k ≤ m/2 do begin
        k := SPARSIFY(P);
        (P^(1), P^(2), ..., P^(k)) := SPLIT(P, k);
        for each q, 1 ≤ q ≤ k, do in parallel
            Compute-by-Splitting(P^(q));
        COLLECT(P, k);
    end
```

According to the recurrence relations (*) and (**), the time for computing the witness table using the procedure *Compute-by-Splitting* is $O(\log m)$ on a CREW PRAM, and $O(\log \log m)$ on a CRCW PRAM. Implementations of the sub-procedures $SPARSIFY$ and $COLLECT$ on a hypercube and on a mesh-connected computer give the results stated in points 3 and 4 above. The constant-time randomized algorithm (point 2) is much more complicated to design. After achieving a large sparsification, the algorithm stops further calls, and begins a special randomized iteration. At this stage, it is more convenient to consider an iterative algorithm. The definition of $SPARSIFY(P)$ needs to be slightly changed: the new version sparsifies the set S of non-computed entries of the witness table assuming that S is already sparse. The basic point is that the sparsity grows according to the inequality:

$$k' \geq k.(m/k)^{1/2},$$

in which k is the old sparsity, and k' is the new sparsity of the set S of non-computed entries. The randomization occurs when sparsity $\geq m^{7/8}$. This is achieved after at most three deterministic iterations.

The constant-time string matching also requires a quite technical (though very interesting) construction of several deterministic samples in constant time. But this is outside the scope of this book. We refer the reader to the bibliographic notes for details.

16.4 Parallel *KMR* algorithm and application

The doubling technique is the crucial feature of the structure of the Karp-Miller-Rosenberg algorithm. It is so again in a parallel setting. At one stage, the algorithm computes the names for all words of size k. At the next stage, using these names, it computes names of words having a size twice as large. We now explain how the algorithm *KMR* of Chapter 7 can be parallelized.

To make a parallel version of *KMR* algorithm, it is sufficient to design an efficient parallel version of one stage of the computation, and this essentially reduces to the parallel computation of *Sort-Rename(x)*. If this procedure is implemented in $T(n)$ parallel time with n processors, we then have a parallel version of *KMR* algorithm working in $T(n) \log n$ time also with n processors. This is due to the doubling technique, and the fact that there are only $\log n$ stages. Essentially, the same problems that are solved by a sequential algorithm can be solved by its parallel version in $T(n) \log n$ time.

The time complexity of computing *Sort-Rename(x)* depends heavily on the model of parallel computation used. It is $T(n) = O(\log n)$ without concurrent writes, and it is $T(n) = O(1)$ with concurrent writes. In the latter case, one needs a memory larger than the total number of operations used by what is called a *bulletin board* (auxiliary table with n^2 entries; or, by using some arithmetic tricks, with $n^{1+\varepsilon}$ entries). This looks slightly artificial, but entries of auxiliary memory do not have to be initialized. The details related to the distribution of processors are also very technical in the case of concurrent writes models. Therefore, we present the algorithms of this section using the PRAM model without concurrent writes. This generally increases the time by a logarithmic factor. This logarithmic factor also allows more time for solving technical problems related to the assignment of processors to elements to be processed. The operation $Sort\text{-}Rename(x)$ basically reduces to sorting, and running it in $O(\log n)$ with $O(n)$ processors is possible. Both in KMR algorithm and in the suffix tree construction we have a logarithmic number of phases, each one using $Sort\text{-}Rename$. This implies directly the following fact.

Theorem 16.4 *The dictionary of basic factors and the suffix tree of a text of length n can be constructed in $\log^2(n)$ time with $O(n)$ processors of a CREW PRAM.*

The theorem has many corollaries, since using the dictionary of basic factors and suffix trees we can solve many other problems in $\log^2(n)$ time with $O(n)$ processors, e.g., square testing, construction of string-matching automata, and searching for symmetries. The dictionary of basic factors leads to another efficient construction of suffix trees, though that is not optimal.

Theorem 16.5 *The algorithm Suffix-Trees-by-Refining can be implemented to work in $O(\log^2 n)$ time with $O(n)$ processors of a CREW PRAM.*

To build the suffix tree of a text, a coarse approximation of it is first built. Afterward the tree is refined step by step. We build a series of a logarithmic number of trees $T_n, T_{n/2}, \ldots, T_1$: each successive tree is an approximation of the suffix tree of the text; the key invariant is:

$inv(k)$: for each internal node v of T_k there are no two distinct outgoing edges for which the labels have the same prefix of the length k; the label of the path from the root to leaf i is $text[i..i+n]$; there is no internal node of outdegree one.

Remark. If $inv(1)$ holds, then, the tree T_1 is essentially the suffix tree $\mathcal{ST}(text)$. Just a trivial modification may be needed to delete all #'s padded for technical reasons, but one. Note that the parameter k is always a power of two. This gives the logarithmic number of iterations.

The core of the construction is the procedure $REFINE(k)$ that transforms T_{2k} into T_k. The procedure maintains the invariant: if $inv(2k)$ is satisfied for T_{2k}, then $inv(k)$ holds for T_k after running $REFINE(k)$ on T_{2k}. The correctness (preservation of invariant) of the construction is based on the trivial fact expressed graphically in Figure 16.1. The procedure $REFINE(k)$ consists of two stages:

(1) insertion of new nodes, one per each non-singleton k-equivalence class,

(2) deletion of nodes of outdegree one.

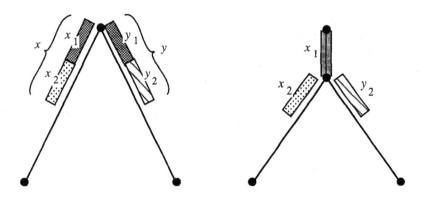

Figure 16.1: If $x \neq y$ and $x_1 = y_1$, then $x_2 \neq y_2$. After insertion of a new node, if $inv(2k)$ is locally satisfied on the left, $inv(k)$ holds locally on the right.

We need the following procedure.

---

procedure *REFINE(k)*;
    for each internal node v of T **do** *LocalRefine(k, v)*;
    delete all nodes of outdegree one;

The informal description of the construction of the suffix tree $\mathcal{ST}(text)$ is summarized by the algorithm below.

Algorithm *Suffix-Tree-by-Refining*;
    let T be the tree of height 1 which leaves are 1, 2, ... ,n,
    the label of the i-th edge is $text[i..n]$ encoded as $[i, *]$;
    $k := n$;
    repeat { T satisfies $inv(k)$ }
        $k := k/2$; *REFINE(k)*;
    until $k = 1$;

In the first stage the operation *LocalRefine(k, v)* is applied to all internal nodes v of the current tree. This local operation is graphically presented in Figures 16.1 and 16.2. The k-equivalence classes, labels of edges outgoing a given node, are computed. For each non-singleton class, we insert a new (internal) node. The algorithm is informally presented on the example text $abaabbaa\#$,

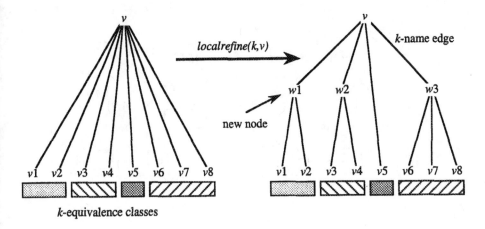

Figure 16.2: Local refinement. The sons of node v for which the edge labels have the same k-prefixes are k-equivalent.

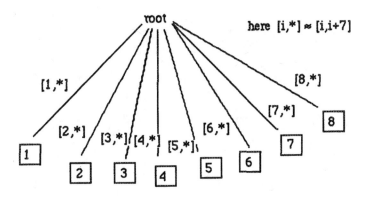

Figure 16.3: The tree T_8 for *text* = *abaabbaa#*. The 8-equivalence is the 4-equivalence, hence $T_8 = T_4$. But the 2-equivalence classes of nodes are $\{2, 6\}, \{3, 7\}, \{1, 4\}, \{8\}, \{5\}$. We apply $REFINE(2)$ to get T_2 (Figure 16.4).

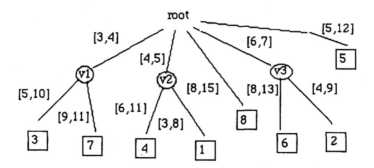

Figure 16.4: The tree T_2. Three new nodes have to be created to obtain T_1. Now the 1-equivalence classes are: $\{3\}, \{7\}, \{4\}, \{1\}, \{v_1, v_2, 8\}, \{6, 2\}, \{v_3, 5\}$.

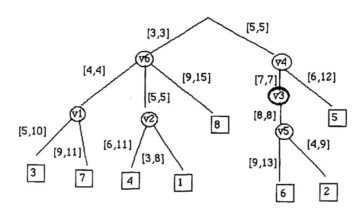

Figure 16.5: Tree T_1 after the first stage of $REFINE(1)$: insertion of new nodes v_4, v_5, v_6.

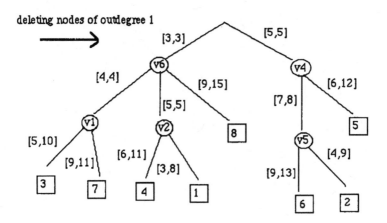

Figure 16.6: Tree T_1 after the second stage of $REFINE(1)$: deletion of node of out-degree one.

see Figure 16.3. We start with the tree T_8 of all factors of length 8 starting at positions $1, 2, \ldots, 8$. The tree is almost a suffix tree: the only condition that is violated is that the root has two distinct outgoing edges in which labels have a common non-empty prefix. We attempt to satisfy the condition by successive refinements: the prefixes violating the condition become smaller and smaller, divided by two at each stage, until they become empty. This is illustrated in the series of Figures 16.4, 16.5, and 16.6.

16.5 Parallel Huffman coding

The sequential algorithm for Huffman coding is quite simple, but unfortunately it appears to be inherently sequential (see Chapter 10). Its parallel counterpart is much more complicated, and requires a new approach. The global structure of Huffman trees must be explored in depth. In this section, we give a poly-logarithmic parallel-time algorithm to compute a Huffman code. The number of processors is $M(n)$, where $M(n)$ is the number of processors needed for a $(\min, +)$ multiplication of two $n \times n$ real matrices in logarithmic parallel time. We assume, for simplicity, that the alphabet is binary.

A binary tree T is said to be *left-justified* if it satisfies the following properties:

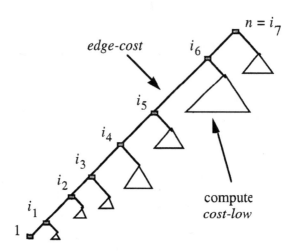

Figure 16.7: A left-justified Huffman tree. The hanging subtrees are of logarithmic height.

1. the depths of the leaves are in non-increasing order from left to right,

2. if a node v has only one son, then it is a left son,

3. let u be a left brother of v, and assume that the height of the subtree rooted at v is at least l. Then the tree rooted at u is full at level l, which means that u has 2^l descendants at distance l.

Most important for the present problem is the following property of left-adjusted trees.

Basic property. Let T be a left-justified binary tree. Then, it has a structure as illustrated in Figure 16.7. All hanging subtrees have height at most $\log n$.

Lemma 16.5 *Assume that the weights p_1, p_2, ... , p_n are pairwise distinct and in increasing order. Then, there is Huffman tree for $(p_1, p_2, \ldots, p_n)$ that is left-justified.*

   Proof. We first show the following claim:

   For each tree T we can find a tree T' satisfying properties $(1), (2), (3)$, in which the leaves are a permutation of leaves of T, and such that the depths of corresponding leaves in the trees T and T' are the same.

The claim can be proven by induction with respect to the height h of the tree T. Let T_1 be derived from T by the following transformation: cut all leaves of T at maximal level; the fathers of these leaves become new leaves, called special leaves in T_1. The tree T_1 has height $h - 1$. It can be transformed into a tree T_1' satisfying (1), by applying the inductive assumption. The leaves of height $h - 1$ of T_1' form a segment of consecutive leaves from left to right. It contains all special leaves. We can permute the leaves at height $h - 1$ in such a way that special leaves are at the left. Then, we insert the deleted leaves back as sons of their original fathers (special leaves in T_1 and T_1'). The resulting tree T' satisfies the claim.

Let us consider a Huffman tree T. It can be transformed into a tree satisfying conditions (1) to (3), with the weight of the tree left unchanged. Hence, after the transformation, the tree is also of minimal weight. Therefore, we can consider that our tree T is optimal *and* is left-justified. It is sufficient to prove that leaves are in increasing order of their weights p_i, from left to right. But this is straightforward since the deepest leaf has the smallest weight. Hence, the tree T satisfies all requirements. □

Theorem 16.6 *The weight of a Huffman tree can be computed in $O(\log^2 n)$ time with n^3 processors. The corresponding tree can be constructed within the same complexity bounds.*

   Proof. Let $weight[i,j] = p_{i+1} + p_{i+2} + \ldots + p_j$. Let $cost[i,j]$ be the weight of a Huffman tree for $(p_{i+1}, p_{i+2}, \ldots, p_j)$ in which the leaves are keys $K_{i+1}, K_{i+2}, \ldots, K_j$. Then, for $i + 1 < j$, we have:

$$(*)cost[i,j] = \min\{cost[i,k] + cost[k,j] + weight[i,j] : i < k < j\}.$$

Let us use the structure of T illustrated in Figure 16.7. All hanging subtrees are shallow; they are of height at most $\log n$. Therefore, we first compute the weights of such shallow subtrees.

Let $cost\text{-}low[i,j]$ be the weight of the Huffman tree of logarithmic height, in which the leaves are keys $K_{i+1}, K_{i+2}, \ldots, K_j$. The table $cost\text{-}low$ can be easily computed in parallel by applying $(*)$. We initialize $cost\text{-}low[i, i+1]$ to p_i, and $cost\text{-}low[i,j]$ to ∞, for all other entries. Then, we repeat $\log n$ times the same parallel-do statement:
  for each i and j, $i < j - 1$, **do in parallel**
    $cost\text{-}low[i,j] := \min\{cost\text{-}low[i,k] + cost\text{-}low[k,j] : i < k < j\}$
        $+ weight[i,j]$.

We need n processors for each operation "min" concerning a fixed pair (i,j). Since there are n^2 pairs (i,j), globally we use a cubic number of processors to compute cost-low's. Now, we have to find the weight of an optimal decomposition of the entire tree T into a leftmost branch, and hanging subtrees. The

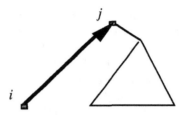

Figure 16.8: $edge\text{-}cost[i,j] = cost\text{-}low[i,j] + weight[i,j]$

consecutive points of this branch correspond to points $(1, i)$. Consider the edge from $(1, i)$ to $(1, j)$, and identify it with the edge (i, j). The contribution of this edge to the total weight is illustrated in Figure 16.8.

We assign to the edge (i, j) the cost given by the formula:

$$edge\text{-}cost(i,j) = cost\text{-}low[i,j] + weight[1,j].$$

It is easy to deduce the following fact: the total weight of T is the sum of costs of edges corresponding to the leftmost branch.

Once we have computed all cost-low's we can assign the weights to the edges according to the formula, and we have an acyclic directed graph with weighted edges. The cost of the Huffman tree is reduced to the computation of the minimal cost from 1 to n in this graph. This can be executed by $\log n$ squaring of the weight matrix of the graph. Each squaring corresponds to a $(\min, +)$ multiplication of $n \times n$ matrices, and, therefore, can be executed in $\log n$ time with n^3 processors. Hence $\log^2 n$ time is sufficient for the entire process. This completes the proof of the first part of the theorem. Given costs, the Huffman tree can be constructed within the same complexity bounds. We refer the reader to [AKLMT 89].   □

In fact matrices that occur in the algorithm have a special property called the *quadrangle inequality* (see Figure 16.9) or Monge property. This property allows the number of processors to be reduced to a quadratic number. A matrix C satisfies the *quadrangle inequality* if for each $i < j < k < l$ we have:

$$C[i, k] + C[j, l] \le C[i, l] + C[j, k].$$

Let us consider matrices that are strictly upper triangular (elements below the main diagonal and on the main diagonal are null). Such matrices correspond to weights of edges in acyclic directed graphs. Denote by Ⓒ the $(\min, +)$ multiplication of matrices:

$$F \textcircled{c} G = C \text{ iff } C[i,j] = \min\{F[i,k] + G[k,j] \ : \ i < k < j\}.$$

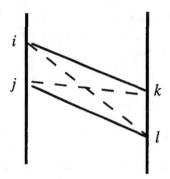

Figure 16.9: The quadrangle inequality: (cost of plain lines) $\leq$ (cost of dashed lines).

The proof of the following fact is left to the reader.

Lemma 16.6 *If matrices F and G satisfy the quadrangle inequality, then $F\copyright G$ also satisfies this property.*

For matrices occurring in the Huffman tree algorithm, the $(\min, +)$ multiplication is simple in parallel, and the number of processors is reduced by a linear factor, due to the following lemma.

Lemma 16.7 *If matrices F and G satisfy the quadrangle inequality, then $F\copyright G$ can be computed in $O(\log^2 n)$ time with $O(n^2)$ processors.*

Proof. Let us fix j, and denote by $CUT[i]$ the smallest integer k for which the value of $F[i,k] + G[k,j]$ is minimal. The computation of $F\copyright G$ reduces to the computation of vectors CUT for each j. It is sufficient to show that, for j fixed, this vector can be computed in $O(\log^2 n)$ time with n processors. The structure of the algorithm to do it is the following:

> let i_{mid} be the middle of interval $[1, 2, \ldots, n]$; compute $CUT[i_{mid}]$ in $O(\log n)$ time with n processors assuming that the value of CUT is in the whole interval $[1, 2, \ldots, n]$.

Afterward, it is easy to see, due to the quadrangle inequality, that

$$CUT[i] \leq CUT[i_{mid}] \text{ for all } i \leq i_{mid},$$
$$CUT[i] \geq CUT[i_{mid}] \text{ for all } i \geq i_{mid}.$$

Using this information we compute $CUT[i]$ for all $i \leq i_{mid}$, knowing that its value is above $CUT[i_{mid}]$. Simultaneously we compute $CUT[i]$ for all $i > i_{mid}$, knowing that its value is not above $CUT[i_{mid}]$. Let m be the size of the interval in which we expect to find the value of $CUT[i]$ (for i in the interval of size n). We have the following equation for the number $P(n, m)$ of processors:

$$P(n, m) \leq \max\{m, P(n, m_1) + P(n, m_2)\}, \text{ where } m_1 + m_2 = m.$$

Obviously $P(n, n) = O(n)$. The depth of the recursion is logarithmic. Each evaluation of a minimum also takes logarithmic time. Hence, we get the required total time. For a fixed j, $P(n, n)$ processors are adequate. Altogether, for all j, we need only a quadratic number of processors.   □

The lemma implies that the parallel Huffman coding problem can indeed be solved in polylogarithmic time with only a quadratic number of processors. This is not optimal, since the sequential algorithm makes only $O(n \log n)$ operations. We refer the reader to [AKLMT 89] for an optimal algorithm.

16.6   Edit distance — efficient parallel computation

The edit distance can be viewed as a shortest path problem on a *grid graph* G (see Chapter 12. For simplicity assume that words x and y are of the same length n. Then, G has $(n + 1)^2$ nodes. Let W be the matrix of weights associated to edges of G. We can use $(\min, +)$ matrix multiplication to obtain the required value of the edit distance. Assume that the weight from the sink node to itself is zero. Then $edit(x, y) = W^n[0, n]$.

The matrix W^n can be computed using successive squaring (or an adaptation of it, if n is not a power of 2):

$$\textbf{repeat } \log n \textbf{ times } W := W^2.$$

Obviously, k^3 processors are sufficient for multiplying two $k \times k$ matrices in $O(\log k)$ time on a CREW PRAM. In our case $k = (n+1)^2$, so, this proves that n^6 processors suffice to compute the edit distance. The time of computation is $O(\log^2 n)$. However, there is a more efficient algorithm, due to the special structure of the grid graph G. The grid graph can be decomposed into four grid graphs of the same type (see Figure 16.10). The partition of G leads to a kind of parallel divide-and-conquer computation.

Theorem 16.7 *The edit distance can be computed in* $\log^3 n$ *time with* n^2 *processors.*

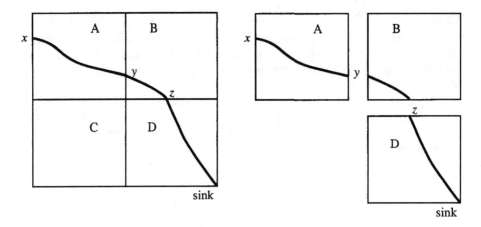

Figure 16.10: Decomposition of the shortest path problem into sub-problems.

Proof. The edit distance problem reduces to the computation of a shortest path from the source (left upper corner) to the sink (right lower corner) on the weighted grid graph G. We compute (recursively) the matrix W of all costs of paths, from positions x on the left or top boundary to any position y on the right or bottom boundary of G. Let us partition the grid into four identical sub-grids, as in Figure 16.10. If we know the matrices of costs between boundary points for all sub-grids, then all costs of paths between boundary positions of the whole grid can be computed with $O(M(n))$ processors in $O(\log^2 n)$ time, using a constant number of matrix multiplications. Here, $M(n)$ denotes the number of processors needed to multiply, in $O(\log^2 n)$ time, two matrices satisfying the quadrangle inequality. We have $M(n) = O(n^2)$. For $n/2 \times n/2$ sub-grids we need $M(n/2)$ processors at one level of recursion. Altogether $M(n)$ processors suffice, since we have the inequality $4M(n/2) \leq M(n)$ (because $M(n) = c\,n^2$). □

Bibliographic notes

The first optimal parallel algorithm for string matching was presented by Galil in [Ga 85]. The algorithm was optimal only for alphabets of constant size. Vishkin improved on the notion of the slow duels of Galil, and described the more powerful concept of (fast) duels that leads to an optimal algorithm independently of the size of the alphabet [Vi 85]. The optimal parallel string-matching algorithm working in $O(\log^2 n)$ time on a CREW PRAM is also from

Vishkin [Vi 85]. The idea of witnesses and duels is used by Vishkin in [Vi 91] in the string matching by sampling. The concept of deterministic sampling is very powerful. It has been used by Galil to design a constant-time optimal parallel searching algorithm (the preprocessing is not included). This result was an improvement upon the $O(\log^* n)$ result of Vishkin, though $O(\log^* n)$ time can also be treated practically as a constant time. The construction of suffix tree by refining was presented originally in [AILSV 88]. The parallel algorithm computing the edit distance is from [AALF 88]. It was also observed independently by Rytter [Ry 88].

Chapter 17

Miscellaneous

This chapter addresses several interesting questions about strings that have not been considered yet in previous chapters: *string-matching by hashing, shortest common superstrings, unique decipherability* problem of codes, *parameterized pattern-matching* and breaking paragraphs into lines. The treatment of problems is not always handled in full details.

17.1   Karp-Rabin string matching by hashing

When we need to compare two objects x and y, we can look at their "fingerprints" given by $hash(x)$, $hash(y)$. If the fingerprints of two objects are equal, there is a strong likelihood that they are really the same object, and we can then apply a more thorough test of equality (if necessary). The two basic properties of fingerprints are:

- efficiently computable

- highly discriminating: it is unlikely to have both $x \neq y$ and $hash(x) = hash(y)$.

The idea of hashing is utilized in the Karp-Rabin string-matching algorithm. The fingerprint FP of the pattern (of length m) is computed first. Then, for each position i on the text, the fingerprint FT of $text[i+1 .. i+m]$ is computed. If ever $FT = FP$, we check directly to see if the equation $pat = text[i+1 .. i+m]$ really holds.

Going efficiently from a position i on the text to the next position $i+1$ requires another property of hashing functions for this specific problem (see Figure 17.1):

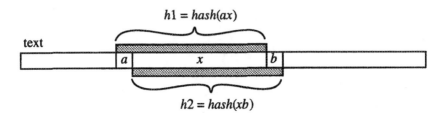

Figure 17.1: $h2 = f(a, b, h1)$, where f is easy to compute.

- $hash(text[i+1..i+m])$ should be easily computable from $hash(text[i..i+m-1])$.

Assume for simplicity that the alphabet is $\{0, 1\}$. Then each string x of length m can be treated as the binary representation of an integer. If m is large, the number becomes too large to fit into a unique memory cell. It is convenient then to take as a fingerprint the value x modulo Q, in which Q is a prime number as large as possible (for example, the largest prime number that fits into a memory cell). The fingerprint function is then

$$hash(x) = [x]_2 \mod Q$$

where $[u]_2$ is the number which binary representation is the string u of length m. All string arguments of $hash$ are of length m. Let $g = 2^{m-1} \mod Q$. Then, the function f (see Figure 17.1) can be computed by the formula:

$$f(a, b, h) = 2(h - ag) + b.$$

Proceeding in this way, the third basic property of fingerprints is satisfied: f is easily computable.

Algorithm *Karp-Rabin*;
    { string-matching by hashing }
    $FP := [pat[1..m]]_2 \mod Q$; $g := 2^{m-1} \mod Q$;
    $FT := [text[1..m]]_2 \mod Q$;
    for $i := 0$ **to** $n - m$ **do begin**
        if $FT = FP$ { small probability } **then**
            check equality $pat = text[i+1..i+m]$
            applying symbol by symbol comparisons
            and report a possible match;
        $FT := f(text[i+1], text[i+m+1], FT)$;
    end

The worst-case time complexity of the algorithm is quadratic. But it could be difficult to find interesting input data causing the algorithm to make effectively a quadratic number of comparisons (the non-interesting example is that in which *pat* and *text* consist only of repetitions of the same symbol). On the average the algorithm is fast, but the best time complexity is still linear. This is to be compared with the lower bound of string matching on the average (which is $O(n \log m/m)$), and the best time complexity of Boyer-Moore type algorithms (which is $O(n/m)$). String matching by hashing produces a straightforward $O(\log n)$ randomized optimal parallel algorithm because the process reduces to prefix computations.

One can also apply other hashing functions that satisfy the three basic properties above. The original *Karp-Rabin* algorithm chooses the prime number randomly.

Essentially, the idea of hashing can also be used to solve the problem of finding repetitions in strings and arrays (looking for repetitions of fingerprints). The algorithm below is an extension of *Karp-Rabin* algorithm to two-dimensional pattern matching. Let m be the number of rows of the pattern array. Fingerprints are computed for columns of the pattern, and for factors of length m of columns of the text array. The problem then reduces to ordinary string matching on the alphabet of fingerprints. Since this alphabet is large, an algorithm in which the performance is independent of the alphabet is actually required.

Algorithm *2D-pattern matching by hashing*;
    { *PAT* is an $m \times m'$ array, T is an $n \times n'$ array }
    $pat := hash(P_1) \ldots hash(P_{m'})$,
        where P_j is the j-th column of *PAT*;
    $text := hash(T_1) \ldots hash(T_{n'})$,
        where T_j is the prefix of length m of the j-th column of T;
    for $i := 0$ **to** $n - m$ **do begin**
        if *pat* occurs in *text* at position j **then**
            check if *PAT* occurs in T at position (i, j)
            applying symbol by symbol comparisons, { cost $O(mm')$ }
            and report a possible match;
        if $i \neq n - m$ **then** { shift one row down }
            for $j := 1$ **to** n' **do**
                $text[j] := f(T[i + 1, j], T[i + m + 1, j], text[j])$;
    end

17.2   Shortest common superstrings

The shortest common superstring problem (SCS) is defined as follows: given a finite set of strings R find a shortest text w such that $R \subseteq \mathcal{F}(w)$. The size of the problem is the total size of all words in R. The superstring w represents in a certain sense all subwords of R.

The problem is known to be NP-complete. So the natural question is to find a polynomial-time approximation for it. The aim of Gallant's method is to compute an approximate solution. The resulting algorithm is called here *Greedy-SCS*. Without loss of generality, we assume (throughout this section) that R is a factor-free set, which means that no word in R is a factor of another word in R. Otherwise, if u is a factor of v $(u, v \in R)$, a solution for $R - \{u\}$ is a solution for R. For two words x and y let us define $Overlap(x,y)$ as the longest prefix of y which is a suffix of x. If $v = Overlap(x,y)$, then x, y are in the form

$$x = u_1 v \text{ and } y = v u_2.$$

Let us define $x \copyright y$ as the word $u_1 v u_2$ $(= u_1 y = x u_2)$. Observe that the shortest superstring of two words u, v is either $u \copyright v$ or $v \copyright u$. Since the set R is factor-free, the operation $\copyright$ has the following properties:

(∗) operation $\copyright$ is associative (on R),

(∗∗) the shortest superstring for R is of the form $x_1 \copyright x_2 \copyright x_3 \copyright \ldots \copyright x_k$, where $x_1, x_2, x_3, \ldots, x_k$ is a permutation of all words of the set R.

---

function *Greedy-SCS(R)*;
    { Gallant's algorithm, greedy approach to SCS }
    if R consists of one word w **then return** w
    else begin
        find two words x, y such that $|Overlap(x,y)|$ is maximal;
        return *Greedy-SCS*$(R - \{x,y\} \cup \{x \copyright y\})$;
    end

The above algorithm is quite effective in term of compression. Let n be the sum of lengths of all words in R. Let w_{min} be a shortest common superstring, and let w_G be the output of Gallant algorithm. Note that $|w_{min}| \leq |w_G| \leq n$. The difference $n - |w_{min}|$ is the size of compression. The smaller the shortest superstring, the better the compression. The following lemma states that the compression reached by Gallant algorithm is at least half the optimal value (see bibliographic notes).

Lemma 17.1 $n - |w_G| \geq (n - |w_{\min}|)/2$.

If we take $w_1 = ab^n$, $w_2 = b^n a$, $w_3 = b^{n+1}$ then the size of the compression is approximately twice the optimal one, if the algorithm first merges w_1 and w_2. Hence the order is essential, unfortunately there generally exponentially many different orderings possible.

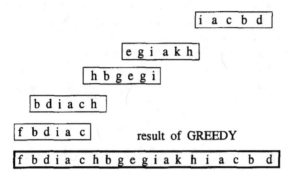

Figure 17.2: The action of the algorithm *Greedy-SCS* on the example strings.

Example. Take the following example set $R = \{w_1, w_2, w_3, w_4, w_5\}$, in which

$$w_1 = egiakh, \ w_2 = fbdiac, \ w_3 = hbgegi, \ w_4 = iacbd, \ w_5 = bdiach.$$

Gallant algorithm produces the following string, see Figure 17.2.

$$
\begin{aligned}
w_G &= \ Greedy\text{-}SCS(w_1, \ w_2\textcircled{c}w_5, \ w_3, \ w_4) \\
&= \ Greedy\text{-}SCS(w_3\textcircled{c}w_1, \ w_2\textcircled{c}w_5, \ w_4) \\
&= \ Greedy\text{-}SCS(w_2\textcircled{c}w_5\textcircled{c}w_3\textcircled{c}w_1, \ w_4) \\
&= \ w_2\textcircled{c}w_5\textcircled{c}w_3\textcircled{c}w_1\textcircled{c}w_4.
\end{aligned}
$$

Its size ℓ is

$$
\begin{aligned}
\ell &= \ n - Overlap(w_2, w_5) - Overlap(w_5, w_3) \\
&\qquad - Overlap(w_3, w_1) - Overlap(w_1, w_4) \\
&= \ 29 - 5 - 1 - 3 - 0 \\
&= \ 20.
\end{aligned}
$$

In this case, $n - |w_G| = 9$.

An alternative approach to *Greedy-SCS* algorithm is to find a permutation

$$x_1, \ x_2, \ x_3, \ \ldots, \ x_k$$

of all words of R, such that $x_1 \copyright x_2 \copyright x_3 \copyright \ldots \copyright x_k$ is of minimal size. This produces exactly a shortest superstring (property **). Translated into graph notation (with nodes x_i's, linked by edges weighted by lengths of overlaps) the problem reduces to the Traveling Salesman Problem, which is also NP-complete. Heuristics for this latter problem can be used for the shortest superstring problem.

The complexity of Gallant algorithm depends on its implementation. Obviously the basic operation is computing overlaps. It is easy to see that for two given strings u and v the overlap is the size of the border of the word $v\#u$. Hence, methods from Chapter 3 (to compute the border table) can be used here. This leads to an $O(nk)$ implementation of Gallant method. The best known implementations of the method work in $O(n \log n)$ time, using sophisticated data structures.

17.3 Unique-decipherability problem

A set of words H, is said to be a *uniquely-decipherable code* if words that are compositions of words of H have only one factorization according to H. The unique decipherability problem consists in testing whether a set of words satisfies the condition. The size n of the problem, when H is a finite set, is the total length of all elements of H; in particular, the cardinality of H is also bounded by n. Note that we can consider that H does not contain the empty word, because otherwise H is not uniquely decipherable and the problem is solved.

Another way to set up the problem is to consider a *coding function* h, or substitution, from B^* to A^* (B and A are two finite alphabets). The function is a morphism (i.e., it satisfies both properties $h(\varepsilon) = \varepsilon$ and $h(uv) = h(u)h(v)$ for all u, v), and the set H, called the code, is $\{h(a) : a \in B\}$. The elements of H are called codewords. Then, asking whether h is a one-to-one function is equivalent to the unique decipherability for H provided all $h(a)$'s are pairwise distinct. Coding functions related to data compression algorithms are considered in Chapter 10. We can assume that all codewords $h(a)$ are non-empty and pairwise distinct. If not, then obviously the function is not one-to-one and the problem is solved.

We translate the unique decipherability condition for H into a problem on a graph G that is now defined. The nodes of G are suffixes of the codewords (including the empty word). There is an edge in G from u to v iff $v = u//x$ or $v = x//u$ for some codeword x. The operation $y//z$ is defined only if z is a prefix of y, that is, if $y = zw$ for some word w, and the result is precisely this word w.

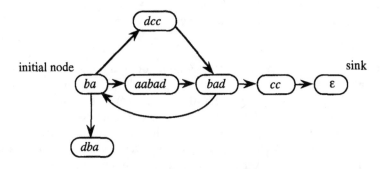

Figure 17.3: Graph G for the example code.

A set of initial nodes, *Init*, is defined for the graph G. Initial nodes are those of the form $x//y$ in which x, y are two distinct codewords. Let us call the empty word the sink. Then, it is easy to prove the following fact:

> the code H is uniquely decipherable iff there is no path in G from an initial node to the sink.

Example. Let $H = \{ab, abba, baaabad, aa, badcc, cc, dccbad, badba\}$. The corresponding graph G is displayed in Figure 17.3. The set is not a uniquely decipherable code because there is a path from ba to the sink. The word ba is an initial node because $ba = abba//ab$. The path is:

$$ba \to aabad \to bad \to cc \to \varepsilon$$

associated with the four equalities $baaabad//ba = aabad$, $aabad//aa = bad$, $badcc//bad = cc$, $cc//cc = \varepsilon$. The path corresponds to two factorizations of the word $abbaaabadcc$:

$$ab.baaabad.cc \text{ and } abba.aa.badcc.$$

The size of graph G is $O(n^2)$, and we can search a path from an initial node to the sink in time proportional to the size of the graph by any standard algorithm. We show that the construction of G can also be accomplished within the same time bound. If we can answer questions like "is $y//x$ defined?" in constant time, then, there is at most a quadratic number of such questions, and we are done. The question "is $y//x$ defined?" is equivalent to "is the x a prefix of y." For a fixed codeword y this takes $O(n + |y|)$ time, since n is the total size of all codewords. Altogether this takes $O(n^2)$ time if we sum over all y's. This proves the following.

Theorem 17.1 *The unique decipherability problem can be solved in $O(n^2)$ time.*

A more precise estimation of the same algorithm shows that it works in $O(nk)$ time, where k is the number of codewords. By applying some data structures, the space complexity can also be improved for some instances of the problem. The close relationship between the unique decipherability problem and accessibility problem in graphs is quite inherent, particularly if space complexity is considered. Indeed, the problems are mutually reducible using additional constant-space memory of a random access machine (or $\log n$ deterministic space of a Turing machine).

17.4   Parameterized pattern matching

Assume we search for a pattern $P \in \Sigma_1^*$ of length m in a text $T \in \Sigma_2^*$ of length n and the same symbol can be written differently in P and T but the full correspondence between symbols can be unknown in advance. For some symbols of Σ_1 we know their equivalent symbols in Σ_2 via an injective *partial naming function*

$$\mathcal{N}_{init} : \Sigma_1 \to \Sigma_2.$$

We are to check if there is an injective *full naming function* $\mathcal{N}$ defined for all symbols of Σ_1, which is an extension of $\mathcal{N}_{init}$, and such that the coded pattern $\mathcal{N}(P)$ occurs in T. We ask for all occurrences for which there exists a corresponding function $\mathcal{N}$; the naming function may differ for each occurrence. The symbols for which $\mathcal{N}_{init}$ is not defined are called *unknown symbols*, so the problem is to check how to name them consistently in such a way that the pattern occurs in T. We assume that the alphabets are *enumerated*, i.e. identified with intervals on natural numbers starting at 1.

Example. Assume that initially $\mathcal{N}_{init}(a) = b$ and $\mathcal{N}$ is not defined for other symbols, consider the following text and pattern:

$$T \;=\; a\,b\,c\,a\,b\,b\,b\,b\,a\,c, \quad P \;=\; a\,b\,c.$$

Then, there are two occurrences of P in T which start at positions 2 and 8. In the first occurrence abc corresponds to bca and in the second one to bac. Hence the naming function for the first occurrence is

$$\mathcal{N} \;:\; a \to b, \; b \to c, \; c \to a$$

and for the second occurrence it is:

$$\mathcal{N} \;:\; a \to b, \; b \to a, \; c \to c.$$

In both correspondences it should be $a \to b$.

We say that u agrees with w and write $u \approx w$ if there is an injective naming function $\mathcal{N}$ such that $\mathcal{N}(u) = w$. Using this terminology we have

$$P \text{ occurs at position } i \text{ in } T \text{ iff } P \approx T[i..i+m-1].$$

Let Σ_0 be the set of symbols for which the initial partial naming function is defined. So Σ_0 is the set of *known symbols* and $A \setminus \Sigma_0$ is the set of *unknown symbols*. For two strings u, w built over the alphabet Σ_1 containing Σ_0 we define the equivalence relation $u \equiv w$ as follows:
(1) u and w have the same length, **(2)** there is a bijection $f : \Sigma_1 \to \Sigma_1$ such that $f(u) = w$ and f is the identity on Σ_0.

Lemma 17.2
(1) If $u \equiv w$ and $w \approx v$ then $u \approx v$.
(2) If $P[1..j] \approx T[i+1..i+j]$ and $1 \le t < j$ then

$$P[1..t] \approx T[i+j-t..i+j] \Rightarrow P[1..t] \equiv P[j-t+1..j].$$

We introduce a new version of the border table which is suitable to using the same strategy for pattern matching with unknown symbols.

For $0 < j \le m$, define

$$ModBord[j] = \max\{0 \le t < j \ : \ P[1..t] \equiv P[j-t+1..j]\}.$$

In other words, the entry $ModBord[j]$ is equal to the length of a longest proper suffix of $P[1..j]$ which *agrees* (with respect to $\equiv$) with a prefix of P of the same length. The entry is null if no non-empty proper suffix satisfy the condition, e.g., it always holds $ModBord[1] = 0$.

In the algorithm we need one table more, called *PRED*. For a string X it is defined by

$$PRED_X[i] = \max\{t < i : X[i] = X[t] \text{ or } t = 0\}$$

for all position i on X. For example, if $X = abcaabac$ we have: $PRED_X = [0, 0, 0, 1, 4, 2, 5, 3]$.

Lemma 17.3 *The table PRED can be computed in linear time.*

Proof. We scan the string X from left to right and use an auxiliary table *LAST*, which records the position of the last occurrence of each symbol before the currently visited position i. Initially *LAST* contains only zeros. And at the i-th iteration we execute the instruction:

$$PRED_X[i] := LAST[X[i]] \; ; \; LAST[X[i]] := i;$$

The computation takes linear time if the alphabet is enumerated.   □

> **function** $MATCH\text{-}EXTENDS(i,j)$;
> { **input:** $P[1..j] \approx T[i+1..i+j]$ and $0 \le j < m$ }
> { **output:** *true* iff $P[1..j+1] \approx T[i+1..i+j+1]$ }
> $u := P[j+1]; \; v := T[i+j+1];$
> **if** $u \in \Sigma_0$ **or** $v \in \mathcal{N}_{init}(\Sigma_0)$ **then**
>   **return** $(\mathcal{N}_{init}(u) = v)$
> **else if** $PRED_P[j+1] > 0$ **then**
>   **return** $(T[i+j+1] = T[i+PRED_P[j+1]])$
> **else**
>   **return** $(PRED_T[i+j+1] \le i)$;

The crucial part of the modified *KMP* algorithm is an auxiliary function *MATCH-EXTENDS*, which checks whether a partial match of $P[1..j]$ against $T[i..i+j]$ extends by one symbol to the right.

Theorem 17.2 [Modified-KMP: searching] *Assume the table ModBord for pattern P is precomputed. Then we can search for P with unknown symbols in a text of length n in time $O(n)$.*

Proof. Algorithm *Modified-KMP* is presented below. It works essentially in the same manner as *KMP* algorithm does; but, instead of making simple symbols comparisons, it uses the special boolean function $MATCH\text{-}EXTENDS(i,j)$. Since running the function *MATCH-EXTENDS* takes only constant time, the algorithm works in linear time. The proof of correctness of the algorithm is less trivial but it follows essentially from Lemma 17.2.   □

> **Algorithm** *Modified-KMP*;
> $i := 0; \; j := 0;$
>   compute tables $PRED_P$ and $PRED_T$;
>   compute table *ModifiedFailureTable*;
>   **while** $i \le n - m$ **do begin**
>   **while** $j < m$ **and** $MATCH\text{-}EXTENDS(i,j)$ **do**
> $j := j+1;$
>   **if** $j = m$ **then** REPORT MATCH at $i+1$;
> $i := i + \max(1, j - ModBord[j]);$
> $j := ModBord[j];$

Theorem 17.3 [Modified-KMP: preprocessing] *The table ModBord for a pattern of length m can be computed in* $O(m)$ *time.*

Proof. We use the algorithm *Computation-of-ModifiedFailureTable*, which is given below. The algorithm mimics a standard algorithm. The function *MATCH-EXTENDS'*(t, k) is a slight modification of *MATCH-EXTENDS*. It checks whether a partial match of $P[1..t]$ against $P[j - t..j]$ extends by one position in the pattern in the sense of relation $\equiv$. The function is designed as a straightforward modification of *MATCH-EXTENDS*. The time complexity of the algorithm *Computation-of-ModifiedFailureTable* is linear due to the fact that *MATCH-EXTENDS'* is computed in constant time. □

Algorithm *Computation-of-ModifiedFailureTable*;
    compute the table $PRED_P$;
    $ModifiedFailureTable[0] := -1$; $t := -1$;
    for $j := 1$ **to** m **do begin**
        while $t \geq 0$ **and not** *MATCH-EXTENDS'*$(t, j - t)$ **do**
            $t := ModifiedFailureTable[t]$;
        $t := t + 1$; $ModifiedFailureTable[j] := t$;

17.5 Breaking paragraphs into lines

In this section, we describe an application of text manipulation to text editing. It is the problem of breaking a paragraph optimally into lines. The algorithm may be seen as another application of the notion of failure function, introduced in Chapter 2 for the *KMP* string-matching algorithm.

The problem of breaking a paragraph is defined as follows. We are given a *paragraph* (a sequence) of n words (in the usual sense) $x_1, x_2, \ldots, x_n$, and bounds *lmin*, *lmax* on lengths of lines. The i-th word of the paragraph has length w_i. A line is an interval $[i..j]$ of consecutive words $x'_k s$ ($i \leq k \leq j$). The length of line $[i..j]$, denoted by $line(i, j)$, is the total length of its words, that is, $w_i + w_{i+1} + \cdots + w_j$. Bounds *lmin* and *lmax* are related to the smallest and largest lengths of lines respectively. The optimal length of a line is *lmax*. Moreover, the length of the line $[i..j]$ is said to be legal if $lmin \leq line(i, j) \leq lmax$. Let us denote the corresponding predicate by $legal(i, j)$.

For a legal line, its penalty is defined as $penalty(i, j) = C.(lmax - line(i, j))$, for some constant C. The problem of breaking a paragraph consists in finding a sequence of integers

$$i_1(= 1), i_2, \ldots, i_k(= n)$$

such that both, lines $[i_1..i_2]$, $[i_2 + 1..i_3]$, ... , $[i_{k-1} + 1..i_k]$ have legal lengths, and the total penalty (sum of penalties of lines) is minimum. Integers i_1, i_2, i_3, ... , i_k are called the breaking points of the paragraph. We assume that there is no penalty (or zero penalty) for the first line, if its length does not exceed *lmax*. It is as if we treated the paragraph starting from the end. In the common definition of the problem the last line is not penalized for being too short. Reversing the order of words in paragraphs leads to an algorithm that is even more similar to *KMP* algorithm, since indices are processed from left to right.

Let *break*[i] be the rightmost breaking point preceding i in an optimal breaking of the subparagraph $[1..i]$. When *break*[i] is computed for all values of i (from 1 to n), the problem is solved: the sequence of breaking points can be recovered by iterating *break* from n.

We first design a brute-force breaking algorithm that runs in quadratic time. The informal scheme of such a naive algorithm is given below. It uses the table f: $f[i]$ is the total penalty of breaking into lines the subparagraph $[1..i]$. The scheme assumes that $f[i]$ is initialized to 0 for all integers i such that $line(1, i) \le lmax$, and is initialized to infinity for all other values of i.

```
for i := 1 to n do begin
    j := an integer which minimizes f[j] + penalty(j + 1, i);
    break[i] := j;
    f[i] := f[j] + penalty(j + 1, i);
end
```

The value of the variable j at the second line is computed by scanning the interval $[first[i], ... , last[i]]$, in which *first*[i] is the smallest integer k for which $legal(k, i)$ holds, and, similarly, *last*[i] is the largest such k less than i. The interval $[first[i], ... , last[i]]$ is called the *legal interval* of i. All values *first*[i], *last*[i], $line(1, i)$ can be precomputed. Therefore, the above scheme yields an $O(n^2)$ time algorithm. The next theorem shows that this can be improved upon considerably.

```
procedure Update-table-Next,
    repeat
        pop the top index j of the stack S; Next[j] := i;
    until g[j] < g[i];
    push i onto S;
```

Algorithm *HL*; { breaking a paragraph of n words into lines }
   precompute values *first*$[i]$, *last*$[i]$, *line*$(1, i)$ for all $i \leq n$;
   let k be the maximal index such that $\text{line}[1, k] \leq lmax$;
   initialize $f[i]$ to 0 and *break*$[i]$ to 0 for all $i \leq k$
      and compute the corresponding values $g[i]$ and *Next*$[i]$;
   $j := 0$;
   for $i := k + 1$ **to** n **do begin**
      { invariant: *break*$[i] < i$, *first*$[i] < $ *last*$[i] < i$ }
      { legal interval is $[first[i]..last[i]]$ }
      if $j < first[i]$ **then** $j := first[i]$;
      while *Next*$[j]$ **defined and** *Next*$[j] \leq last[i]$ **do**
         {j is not rightmost minimal } $j := Next[j]$;
      break$[i] := j$; $f[i] := f[j] + penalty(j + 1, i)$;
      $g[i] := f[i] + C.line(1, i)$;
      Update-table-next;
   end;
   return table *break*, which gives an optimal breaking;

Theorem 17.4 *The problem of optimally breaking a paragraph of n words can be solved in $O(n)$ time.*

Proof. Let us define the function g by $g[j] = f[j] + C.line(j, n)$. The crucial point is the following property: a value of j that minimizes $f[j] + penalty(j + 1, i)$ also minimizes $g[j]$ in the legal interval of j.

That the difference between expressions depends only on i can be checked using simple arithmetics. Another important property is the monotonicity of breaking points:

$$i' < i \implies break[i'] \leq break[i].$$

Hence, the value of j is to be found in the interval $[\max(first[i], break[i - 1])..last[i]]$. Define *Next*$[j]$ to be the first position $k \leq i$ to the right of j such that $g[k] \leq g[j]$. When looking for the minimal value of $g[j]$ in a legal interval, we can initialize j to the beginning of the interval, and compute successive positions by iterating *Next*:

$$j_1 = Next[j], j_2 = Next[j_1], \ldots$$

until the value is undefined, or until it goes outside the interval. Doing so, *Next* works as the failure table of *KMP* algorithm. This produces the algorithm presented below. It works in linear time, as per a similar argument as that used in the analysis of *KMP* algorithm, if the total cost of updating the table

Next is linear. Let $j_1, j_2, \ldots, j_r$ be the increasing sequence of values of j for which *Next*$[j]$ is not defined at a given stage of the algorithm. Then the values of $g[j]$ are strictly increasing for this sequence. We keep the sequence $j_1, j_2, \ldots, j_r$ on a stack S, the last element at the top. The procedure *Update-table-Next* is then applied. The total complexity of computing the table *Next* is linear, since each position is popped from stack S at most once. $\Box$

Bibliographic notes

String matching by hashing was first considered by Harrison [Ha 71]. A complete analysis is presented by Karp and Rabin in [KR 87]. The same idea (using hashing) is applied to finding repetitions in [Ra 85]. An adaptation of *Karp-Rabin* algorithm to two-dimensional pattern matching has been designed by Feng and Takaoka [FT 89]. The approximation of the SCS is from [GMS 80]. An efficient implementation has been designed by Tarhio and Ukkonen [TU 88]. Stronger methods are developed in [Tu 89], providing an $O(n \log n)$-time algorithm, there are plenty of other algorithms for the problem. The algorithm for testing unique decipherability of a code is usually attributed to Sardinas and Paterson (see [Lo 83]). The problem is complete in the class of non-deterministic $\log n$-space computations (see [Ry 86]). The parameterized pattern matching was introduced by Baker [Ba 93]. It has also been considered by Amir, Farach, Muthukrishnan [AFM 94]. We present here our own version. The application of failure functions to the problem of breaking a paragraph into lines is from Hirschberg and Larmore [HL 87a]. There are many stringology subjects which are not covered in the book, one of them is recently developed algorithmics on compressed strings, see [Ry 00], [KS 99], [Ry 02a]. Another problem is the solvability of word equations, the main jewel in this area is the algorithm of Plandowski [Pl 99]. However this problem is almost certainly beyond the class of polynomially solvable problems (the best algorithm works in a polynomial space). There are many NP-hard stringology problems, for example word equations and the superstring problem. However our book was mostly devoted to (deterministic) polynomial-time algorithms.

Bibliography

[Ab 89] Abrahamson, D.M., Generalized string-matching, SIAM J. Comput. 16 (1989): 77–83.

[Ah 80] Aho, A.V., Pattern matching in strings, in (Book, editor, Formal Language Theory — Perspectives and Open Problems, Academic Press, Orlando, Florida 1980): 325–347.

[Ah 90] Aho, A.V., Algorithms for finding patterns in strings, in (J. van Leeuwen, editor, Handbook of Theoretical Computer Science, vol A, Algorithms and complexity, Elsevier, Amsterdam, 1990): 255–300.

[AC 75] Aho, A.V., and Corasick, M., Efficient string matching: An aid to bibliographic search, Comm. ACM 18 (1975): 333–340.

[AHU 76] Aho, A.V., Hirschberg, D.S., and Ullman, J.D., Bounds on the complexity of the longest common subsequence problem, J. ACM 23 (1976): 1–12.

[AHU 74] Aho, A.V., Hopcroft, J.E., and Ullman, J.D., The Design and Analysis of Computer Algorithms, Addison-Wesley, Reading, Mass., 1974.

[AHU 83] Aho, A.V., Hopcroft, J. E., and Ullman, J.D., Data Structures and Algorithms, Addison-Wesley, Reading, Mass., 1983.

[AKW 88] Aho, A.V., Kernighan, B.W. and Weinberger, P.J., The AWK Programming Language, Addison-Wesley, Reading, Mass., 1988.

[ASU 86] Aho, A.V., Sethi, R., and Ullman, J.D., Compilers-Principles, Techniques and Tools, Addison-Wesley, Reading, Mass., 1986.

[AD 86] Allison, L., and Dix, T.I., A bit string longest common subsequence algorithm, Inf. Process. Lett. 23 (1986): 305–310.

[ABF 92a] Amir, A., Benson, G., Two-dimensional periodicity and its application, in (Proc. Symp. On Discrete Algorithms, 1992): 440–452.

[**ABF 92b**] Amir, A., Benson, G., and Farach, M., Optimal parallel two-dimensional pattern matching, in (5th Annual ACM Symposium on Parallel Algorithms and Architectures, ACM Press, 1993) 79–85.

[**ACH 01**] Amir, A., Cole, R., Hariharan, R., Lewenstein, M., and Porat, E., Overlap matching, in: (Twelfth Annual ACM-SIAM Symposium on Data Structures (SODA), 2001).

[**AF 91**] Amir, A., and Farach M., Efficient two-dimensional approximate matching of non-rectangular figures, in (Proc. Symp. On Discrete Algorithm, 1991): 212–223.

[**AFM 94**] Amir, A., Farach, M., and Muthukrishnan, S., Alphabet Dependence in Parameterized Matching, Inf. Process. Lett. 49 (1994) 111–115.

[**AL 91**] Amir, A., and Landau, G.M., Fast parallel and serial multidimensional approximate array matching, Theoret. Comput. Sci. 81 (1991): 97–115.

[**ALV 92**] Amir, A., Landau, G.M., and Vishkin, U., Efficient pattern matching with scaling, J. Algorithms 13 (1992): 2–32.

[**Ap 85**] Apostolico, A., The myriad virtues of suffix trees, in: [AG 85]: 85–96.

[**Ap 86**] Apostolico, A., Improving the worst-case performance of the Hunt Szymanski strategy for the longest common subsequence of two strings, Inf. Process. Lett. 23 (1986): 63–69.

[**Ap 87**] Apostolico, A., Remark on the Hsu-Du new algorithm for the longest common subsequence problem, Inf. Process. Lett. 25 (1987): 235–236.

[**Ap 92**] Apostolico, A., Fast parallel detection of squares in strings, Algorithmica 8 (1992): 285–319.

[**AALF 88**] Apostolico, A., Atallah, M.J., Larmore, L.L., and McFaddin, H.S., Efficient parallel algorithms for string editing and related problems, SIAM J. Comput. 19,5 (1990): 968–988.

[**ABG 92**] Apostolico, A., Breslauer, D., and Galil, Z., Optimal parallel algorithms for periods, palindromes and squares, in: (3rd proceedings of SWAT, LNCS 621, Springer-Verlag, 1992): 296–307.

[**ABG 92**] Apostolico, A., Browne, S., and Guerra, C., Fast linear-space computations of longest common subsequences, Theoret. Comput. Sci. 92 (1992): 3–17.

[**AC 91**] Apostolico, A., and Crochemore, M., Optimal canonization of all substrings of a string, Information and Computation 95, 1 (1991): 76–95.

[**AC 90**] Apostolico, A., and Crochemore, M., Fast parallel Lyndon factorization and applications, Math. Syst. Theory 28,2 (1995): 89–108.

[**AG 85**] Apostolico, A., and Galil, Z., editors, Combinatorial Algorithms on Words, NATO Advanced Science Institutes, Series F, vol 12, Springer-Verlag, Berlin, 1985.

[**AG 97**] Apostolico, A., and Galil, Z., editors, Pattern matching algorithms, Oxford University Press (1997)

[**AG 84**] Apostolico, A., and Giancarlo, R., Pattern-matching machine implementation of a fast test for unique decipherability, Inf. Process. Lett. 18 (1984): 155–158.

[**AG 86**] Apostolico, A., and Giancarlo, R., The Boyer-Moore-Galil string-searching strategies revisited, SIAM J.Comput. 15 (1986): 98–105.

[**AG 87**] Apostolico, A., and Guerra, C., The longest common subsequence problem revisited, J. Algorithms 2 (1987): 315–336.

[**AILSV 88**] Apostolico, A., Iliopoulos, C., Landau, G.M., Schieber, B., and Vishkin, U., Parallel construction of a suffix tree with applications, Algorithmica 3 (1988): 347–365.

[**AP 83**] Apostolico, A., and Preparata, F.P., Optimal off-line detection of repetitions in a string, Theoret. Comput. Sci. 22 (1983): 297–315.

[**AP 85**] Apostolico, A., and Preparata, F.P., Structural properties of the string statistics problem, J. Comput. Syst. Sci. 31 (1985): 394–411.

[**AKLMT 89**] Atallah, M.J., Kosaraju, S.R., Larmore, L.L., Miller, G.L., and Teng, S-H., Constructing trees in parallel, Report CSD-TR-883, Purdue University, 1989.

[**Ba 88**] Baase, S., Computer Algorithms-Introduction to Design and Analysis, Addison-Wesley, Reading, Mass., 1988, 2nd edition.

[**Ba 89**] Baeza-Yates, R.A., Improved string searching, Software Practice and Experience 19 (1989) 257–271.

[**BCG 93**] Baeza-Yates, R.A., Choffrut, C., and Gonnet, G.H., On Boyer-Moore automata, Algorithmica 12,4/5 (1994): 268–292.

[**BG 89**] Baeza-Yates, R.A., and Gonnet, G.H., Efficient text searching of regular expressions, in: (Automata, Languages and Programming, LNCS 372, Springer-Verlag, Berlin, 1989): 46–62.

[**BG 92**] Baeza-Yates, R.A., and Gonnet, G.H., A new approach to text searching, Comm. ACM 35, 10 (1992): 74–82.

[**BGR 90**] Baeza-Yates, R.A., Gonnet, G.H., and Régnier, M., Analysis of Boyer-Moore type string searching algorithms, in: (Proc. of Ist ACM-SIAM Symposium on Discrete Algorithms, American Mathematical Society, Providence, 1990): 328–343.

[BR 90] Baeza-Yates, R.A., and Régnier, M., Fast algorithms for two-dimensional and multiple pattern matching, in (Proc. 2nd Scandinavian Workshop in Algorithms Theory, LNCS 447, Springer-Verlag, Berlin, 1990): 332–347.

[BR 92] Baeza-Yates, R.A., and Régnier, M., Average running time of the Boyer-Moore-Horspool algorithm, Theoret. Comput. Sci. 92 (1992): 19–31.

[Ba 93] Baker, B., A theory of parameterized pattern matching: Algorithms and applications, in: (STOC'93, 1993): 71–80.

[Ba 78] Baker, T.P., A technique for extending rapid exact-match string matching to arrays of more than one dimension, SIAM J.Comput. 7 (1978): 533–541.

[Ba 81] Barth, G., An alternative for the implementation of the Knuth-Morris-Pratt algorithm, Inf. Process. Lett. 13 (1981): 134–137.

[Ba 84] Barth, G., An analytical comparison of two string-searching algorithms, Inf. Process. Lett. 18 (1984): 249–256.

[Ba 85] Barth, G., Relating the average-case cost of the brute-force and the Knuth-Morris-Pratt string-matching algorithm, in [AG 85]: 45–58.

[BEM 79] Bean, D., Ehrenfeucht, A., and Mc Nully, G., Avoidable patterns in strings of symbols, Pacific Journal of Math. 85 (1979): 261–294.

[BBC 92] Beauquier, D., Berstel, J., and Chrétienne, P., Éléments d'Algorithmique, Masson, Paris, 1992.

[BCW 90] Bell, T.C., Cleary, J.G., and Witten, I.H., Text Compression, Prentice Hall, Englewood Cliffs, New Jersey, 1990.

[BSTW 86] Bentley, J.L., Sleator, D.D., Tarjan, R.E., and Wei V.K., A locally adaptive data compression scheme, Commun. ACM 29, 4 (1986): 320–330.

[BBGSV 89] Berkman, O., Breslauer, D., Galil, Z., Schieber, B., and Vishkin, U., Highly parallelizable problems, in: (Proc. 21st ACM Symposium on Theory of Computing, Association for Computing Machinery, New York, 1989): 309–319.

[BP 85] Berstel, J., and Perrin, D., Theory of Codes, Academic Press, Orlando, Florida, 1985.

[Bi 77] Bird, R.S., Two-dimensional pattern matching, Inf. Process. Lett. 6 (1977) 168–170.

[BR 87] Bishop, M.J., and Rawling, C.J., Nucleic Acid and Protein Sequence Analysis: A Practical Approach, IRL Press Limited, Oxford, England, 1987.

[BBEHM 83] Blumer, A., Blumer, J., Ehrenfeucht, A., Haussler, D., and McConnell, R., Linear size finite automata for the set of all subwords of a word an outline of results, Bull. Europ. Assoc. Theoret. Comput. Sci. 21 (1983): 12–20.

[BBEHCS 85] Blumer, A., Blumer, J., Ehrenfeucht, A., Haussier, Chen, M.T., and Seiferas, J., The smallest automaton recognizing the subwords of a text, Theoret. Comput. Sci. 40 (1985): 31–55.

[BBEHM 87] Blumer, A., Blumer, J., Ehrenfeucht, A., Haussier, D., and Mc Connell, R., Complete inverted files for efficient text retrieval and analysis, J. ACM 34 (1987): 578–595.

[BI 87] Blumer, J.A., How much is that dawg in the window? A moving window algorithm for the directed acyclic word graph, J. Algorithms 8 (1987): 451–469.

[Bo 80] Booth. K., Lexicographically least circular strings, Inf. Process. Lett. 10 (1980): 240–242.

[BM 77] Boyer, R.S., and Moore, J.S., A fast string-searching algorithm, Comm. ACM 20 (1977): 762–772.

[BG 90] Breslauer, D., and Gall, Z., An optimal $O(\log\log n)$-time parallel string-matching, SIAM J. Comput 19,6 (1990): 1051–1058.

[BCT 93] Breslauer, D., Colussi, L., and Toniolo, L., Tight comparison bounds for the string prefix-matching problem, Inf. Process. Lett. 47 (1993): 51–57.

[Ca 90] Capocelli, R., editor, Sequences: Combinatorics, Compression, Security and Transmission, Springer-Verlag, New-York, 1990.

[CSV 93] Capocelli, R., and de Santis, A., Vaccaro, U., editors, Sequences II, Springer-Verlag, New-York, 1993.

[CL 90] Chang, W.I., and Lawler, E.L., Approximate string matching in sublinear expected time, in: (FOCS'90): 116–124.

[CL 97] Charras, C., and Lecroq, T., Exact String Matching Algorithms, http://www-igm.univ-mlv.fr/ lecroq/string/, 1997.

[CL 98] Charras, C., and Lecroq, T., Sequence Comparison, http://www-igm.univ-mlv.fr/ lecroq/seqcomp/, 1998.

[CS 85] Chen, M.T., and Seiferas, J., Efficient and elegant subword tree construction, in: [AG 85]: 97–107.

[Ch 90] Choffrut, C., An optimal algorithm for building the Boyer-Moore automaton, Bull. Europ. Assoc. Theoret. Comput. Sci 40 (1990): 217–225.

[CR 87a] Chrobak, M., and Rytter, W., Remarks on string matching and one-way multi-head automata, Inf. Process. Lett. 24 (1987): 325–329.

[CR 87b] Chrobak, M., and Rytter, W., Unique decipherability for partially commutative alphabets, Fundamenta Informaticae (1987): 323–336.

[CS 75] Chvatal, V., and Sankoff, D., Longest common subsequence of two random sequences, J. Appl. Prob. 12 (1975): 306–315.

[Co 88] Cole, R., Parallel merge sort, SIAM J.Comput. 17 (1988): 770–785.

[Co 91] Cole, R., Tight bounds on the complexity of the Boyer-Moore algorithm, in: (Proceedings of the Second Annual ACM-SIAM Symposium on Discrete Algorithms, 1991): 224–233.

[Co 94] Cole, R., Tight bounds on the complexity of the Boyer-Moore string-matching algorithm, SIAM J. Comput. 23,5 (1994): 1075–1091.

[CHI 99] Cole, R., Hariharan, R., and Indyk, P., Tree pattern matching and subset matching in deterministic $O(nlog^3m)$ time, Proceedings of the Tenth Annual ACM-SIAM Symposium on Discrete Algorithms (SODA), 1999, 245-254.

[C-R 93b] Cole, R., Crochemore, M., Galil, Z., Gasieniec, L., Hariharan, R., Muthakrishnan, S., Park, K., and Rytter, W., Optimally fast parallel algorithms for preprocessing and pattern matching in one and two dimensions, in: (FOCS'93, 1993): 248–258.

[CH 92] Cole, R., and Hariharan, R., Tighter bound on the exact complexity of string matching, in: (FOCS'92): 600–609.

[CH99] Cole, R., and Hariharan, R., Faster suffix tree construction with missing suffix links, FOCS 1999.

[CGG 90] Colussi, L., Galil, Z., and Giancarlo, R., On the exact complexity of string matching, in (Proc. 31st Symposium on Foundations of Computer Science, IEEE, 1990): 135–143.

[Co 79] Commentz-Walter, B., A string-matching algorithm fast on the average, in: (Automata, Languages and Programming, Lecture Notes in Computer Science, Springer-Verlag, Berlin, 1979): 118–132.

[CD 89] Consel, C., and Danvy, O., Partial evaluation of pattern matching in strings, Inf. Process. Lett. 30 (1989): 79–86.

[Co 72] Cook, S.A., Linear-time simulation of deterministic two-way push-down automata, Inf. Process. Lett. 71 (1972): 75–80.

[CH 84] Cormack, G.V., and Horspool, R.N.S., Algorithms for adaptive Huffman codes, Inf. Process. Lett. 18 (1984): 159–165.

[CLR 89] Cormen, T.H., Leirserson, C.E., and Rivest, R.L., Introduction to Algorithms, The MIT Press, Cambridge, Mass., 1989.

[Cr 81] Crochemore, M., An optimal algorithm for computing the repetitions in a word, Inf. Process. Lett. 12 (1981): 244–250.

[Cr 83] Crochemore, M., Recherche linéaire d'un carré dans un mot, C. R. Acad. Sc. Paris, t. 296 (1983) Série 1, 781–784.

[Cr 85] Crochemore, M., Optimal factor transducers, in: [AG 85]: 31–43.

[Cr 86] Crochemore, M., Transducers and repetitions, Theoret. Comput. Sci. 45, (1986) 63–86.

[Cr 87] Crochemore, M., Longest common factor of two words, in: (TAP-SOFT'87, vol 1, Springer-Verlag, Berlin, 1987): 26–36.

[Cr 92] Crochemore, M., String matching on ordered alphabets, Theoret. Comput. Sci. 92 (1992): 33–47.

[C-R 92] Crochemore, M., Czumaj, A., Gasieniec, L., Jarominek, S., Lecroq, T., Plandowski, W., and Rytter, W., Speeding up two string-matching algorithms, in: (9th Annual Symposium on Theoretical Aspects of Computer Science, Springer-Verlag, Berlin, 1992): 589–600.

[C-R 93a] Crochemore, M., Czumaj, A., Gasieniec, L., Jarominek, S., Lecroq, T., Plandowski, W., and Rytter, W., Fast multi-pattern matching, Inf. Process. Lett. 71,3–4 (1999) 107–113.

[CGPR93] Crochemore, M., Gasieniec, L., and Rytter, W., Two-dimensional pattern matching by sampling, Inf. Process. Lett. 96 (1993): 159–162.

[CGPR95] Crochemore, M., Gasieniec, L., Plandowski., W., and Rytter, W. Two dimensional pattern matching in linear time and small space, STACS'95.

[CHL 01] Crochemore, M., Hancart, C., and Lecroq, T., Algorithmique du texte, Vuibert Informatique, Paris, 2001.

[CLU 02] Crochemore, M., Landau, G., and Ziv-Ukelson, M., A sub-quadratic sequence alignment algorithm for unrestricted cost matrices, In (Proceedings of the Thirteen Annual ACM-SIAM Symposium on Discrete Algorithms, ACM-SIAM, 2002): 679–688.

[CL 97] Crochemore, M., and Lecroq, T., Tight bounds on the complexity of the Apostolico-Giancarlo algorithm, Information Processing Letters, 63 (1997): 195–203.

[CMRS 00] Crochemore, M., Mignosi, F., Restivo, A., and Salemi, S., Data compression using antidictonaries, Proceedings of the I.E.E.E., 88 (2000):1756–1768.

[CP 91] Crochemore, M., and Perrin, D., Two-way string matching, J. ACM 38, 3 (1991): 651–675.

[CR 90] Crochemore, M., and Rytter, W., Parallel construction of minimal suffix and factor automata, Inf. Process. Lett. 35 (1990): 121–128.

[CR 91a] Crochemore, M., and Rytter, W., Efficient parallel algorithms to test square-freeness and factorize strings, Inf. Process. Lett. 38 (1991): 57–60.

[CR 91c] Crochemore, M., and Rytter, W., Usefulness of the Karp Miller Rosenberg algorithm in parallel computations on strings and arrays, Theoret. Comput. Sci. 88 (1991): 59–82.

[CR 92] Crochemore, M., and Rytter, W., Note on two-dimensional pattern matching by optimal parallel algorithms, in: (Parallel Image Analysis, LNCS 654, Springer-Verlag, 1992): 100–112.

[CR 94] Crochemore, M., and Rytter, W., Text algorithms, Oxford University Press (1994).

[CR95a] Crochemore, M., and Rytter, W. On alphabet-independent linear time algorithms for two-dimensional pattern-matching, in: (LATIN'95, LNCS 911, Springer-Verlag, 1995) 220–229.

[CR 95b] Crochemore, M., and Rytter, W., Squares, cubes and time-space efficient string searching, Algorithmica 13,5 (1995): 405–425.

[De 79] Deken, J., Some limit results for longest common subsequences, Discrete Math. 26 (1979): 17–31.

[DGM 90] Dubiner, M., Galil, Z., and Magen, E., Faster tree-pattern matching, in: (Proceedings of 31st FOCS, 1990): 145–150.

[Du 79] Duval, J.-P., Périodes et répétitions des mots du monoide libre, Theoret. Comput. Sci. 9 (1979): 17–26.

[Du 82] Duval, J.-P., Relationship between the period of a finite word and the length of its unbordered segments, Discrete Math. 40 (1982): 31–44.

[Du 83] Duval, J.-P., Factorizing words over an ordered alphabet, J. Algorithms 4 (1983): 363–381.

[EV 88] Eilam-Tzoreff, T., and Vishkin, U., Matching patterns in strings subject to multi-linear transformations, Theoret. Comput. Sci. 60,3 (1988): 231-254.

[EGGI 92] Eppstein, D., Galil, Z., Giancarlo, R., and Italiano, G., Sparse dynamic programming I, J. ACM 39 (1990): 519–545.

[Fa 73] Faller, N., An adaptive system for data compression, in: (Record of the 7th Asilomar Conference on Cincuits, Systems, and Computers, 1973): 593–597.

[Fa 97] Farach, M., Optimal suffix tree construction with large alphabets, FOCS 1997.

[FFM 00] Farach, M. , Ferragina, P., and Muthukrishnan, S. On the sorting complexity of suffix tree construction, Journal of the ACM, vol. 47(6), 987–1011, 2000.

[FT 87] Feng, Z.R., and Takaoka, T., On improving the average case of the Boyer-Moore string-matching algorithm, J. Inf. Process. 10, 3 (1987): 173–177.

[FT 89] Feng, Z.R., and Takaoka, T., A technique for two-dimensional pattern matching, Comm. ACM 32 (1989): 1110–1120.

[Fe97] Ferragina, P., Dynamic text indexing under string updates, Journal of Algorithms, 22(2):296-328, 1997.

[FG98] Ferragina, P., and Grossi, R., Optimal On-Line Search and Sublinear Time Update in String Matching, SIAM Journal on Computing 27 (1998).

[FW 65] Fine, N.J., and Wilf, H.S., Uniqueness theorems for periodic functions, Proc. Amer. Math. Soc. 16 (1965): 109–114.

[FP 74] Fischer, M.J., and Paterson, M.S., String matching and other products, in:(Proc. of SIAM-AMS Conference on Complexity of Computation, American Mathematical Society, Providence, R.I., 1974) 113–125.

[Fr 75] Fredman, M.L., On computing the length of longest increasing subsequences, Discrete Math. 11 (1975): 29–35.

[Ga 76] Galil, Z., Two fast simulations which imply fast string-matching and palindrome-recognition algorithms, Inf. Process. Lett. 4,4 (1976): 85–87.

[Ga 77] Galil, Z., Some open problems in the theory of computations as questions about two-way deterministic pushdown automaton languages, Math. Syst. Theory 10 (1977): 211–228.

[Ga 78] Galil, Z., Palindrome recognition in real time by a multitape Turing machine, J. Comput. Syst. Sci 16 (1978): 140–157.

[Ga 79] Galil, Z., On improving the worst case-running time of the Boyer-Moore string-searching algorithm, Comm. ACM 22 (1979): 505–508.

[Ga 81] Galil, Z., String matching in real time, J. ACM 28 (1981): 134–149.

[Ga 85a] Galil, Z., Open problems in stringology, in: [AG 85]: 1–12.

[Ga 85b] Galil, Z., Optimal parallel algorithm for string matching, Information and Control 67 (1985): 144–157.

[Ga 92] Galil, Z., A constant-time optimal parallel string-matching algorithm, in: (Proc. 24th ACM Symp. on Theory Of Computing, 1992): 69–76.

[**GG 87**] Galil, Z., and Giancarlo, R., Parallel string matching with k mismatches, Theoret Comput. Sci. 51 (1987) 341–348.

[**GG 88**] Galil, Z., and Giancarlo, R., Data structures and algorithms for approximate string matching, J. Complexity 4 (1988): 33–72.

[**GG 89**] Galil, Z., and Giancarlo, R., Speeding up dynamic programming with applications to molecular biology, Theoret. Comput. Sci. 64 (1989): 107–118.

[**GP 89**] Galil, Z., and Park, K., An improved algorithm for approximate string matching, in: (Automata, Languages and Programming, Lecture Notes in Computer Science 372, Springer-Verlag, Berlin, 1989): 394–404.

[**GP 92a**] Galil, Z., and Park, K., Dynamic programming with convexity, concavity and sparsity, Theoret. Comput. Sci. 92 (1992): 49–76.

[**GP 92b**] Galil, Z., and Park, K., Truly alphabet-independent two-dimensional matching, in: (Proc. 33rd Annual IEEE Symposium on the Foundations of Computer Science, 1992): 247–256.

[**GR 92**] Galil, Z., and Rabani, Y., On the sparse complexity of some algorithms for sequence comparison, Theoret. Comput. Sci. 95, 2 (1992): 231–244.

[**GS 78**] Galil, Z., and Seiferas, J., A linear-time on-line recognition algorithm for 'Palstars, J. ACM 25 (1978): 102–111.

[**GS 80**] Galil, Z., and Seiferas, J., Saving space in fast string matching, SIAM J. Comput. 9 (1980): 417–438.

[**GS 81**] Galil, Z., and Seiferas, J., Linear-time string matching using only a fixed number of local storage locations, Theoret. Comput. Sci. 13 (1981): 331–336.

[**GS 83**] Galil, Z., and Seiferas, J., Time-space optimal string matching, J. Comput. Syst. Sci. 26 (1983): 280–294.

[**Ga 78**] Gallager, R. G. , Variations on a theme by Huffman, I. E. E. E. Trans. Inform. Theory IT 24,6 (1978): 668–674.

[**GMS 80**] Gallant, J., Maier, D., and Storer, J.A., On finding minimal length superstrings, J. Comput. SYSL Sci. 20 (1980): 50–58.

[**GJ 79**] Garey, M.R., and Johnson, D.S., Computers and Intractability: A Guide to the Theory of NP-Completeness, W.H. Freeman, New York, 1979.

[**Gi 93**] Giancarlo, R., The suffix tree of a square matrix, with applications, in: (Proc. Symp. On Discrete Algorithms, 1993): 402–411.

[GR 86] Gibbons, A., and Rytter, W., On the decidability of some problems about rational subsets of free partially commutative monoids , Theoret. Comput. Sci. 48 (1986): 329–337.

[GR 88] Gibbons, A., and Rytter, W., Efficient Parallel Algorithms, Cambridge University Press, Cambridge, U.K., 1988.

[GR 89] Gibbons, A., and Rytter, W., Optimal parallel algorithm for dynamic expression evaluation and application to context-free recognition, Information and Computation 81, 1 (1989): 32–45.

[GB 91] Gonnet, G.H., and Baeza-Yates, R., Handbook of Algorithms and Data Structures, Addison-Wesley, Reading, Mass., 1991.

[Gr 91] Gross, M., Constructing lexicon-grammars, in: (Computational Approaches to the Lexicon, Oxford University Press, 1991).

[GP 89] Gross, M., and Perrin, D., editors, Electronic Dictionaries and Automata in Computational Linguistics, Lecture Notes in Computer Science 377, Springer-Verlag, Berlin, 1989.

[GV 00] Grossi, R., and Vitter, J. S., Compressed Suffix Arrays and Suffix Trees with Applications to Text Indexing and String Matching, ACM Symposium on the Theory of Computing (2000).

[GO 80] Guibas, L.J., and Odlyzko, A.M., A new proof of the linearity of the Boyer-Moore string-searching algorithm, SIAM J. Comput. 9 (1980): 672–682.

[GO 81a] Guibas, L.J., and Odlyzko. A.M Periods in strings, J. Comb. Th. A 30 (1981): 19–42.

[GO 81b] Guibas, L.J., and Odlyzko, A.M., String overlaps, pattern matching and non-transitive games, J. Comb. Th. A 30 (1981): 183–208.

[Gu 97] Gusfield, D., Algorithms on strings, trees and sequences, computer science and computational biology, Cambridge University Press (1997).

[HD 80] Hall, P.A.V., and Dowling, G.R., Approximate string matching, ACM Comput. Surv. 12 (1980): 381–402.

[Ha 93] Hancart, C., On Simon's string-searching algorithm, Inf Process. Lett. 47 (1993): 95–99.

[HPS 92] Hansel, G., Perrin, D., and Simon, I., Compression and entropy, in: (STACS 92, LNCS 577, Springer-Verlag, Berlin, 1992): 515–528.

[HT 84] Harel, D., and Tarjan, R.E., Fast algorithms for finding nearest common ancestors, SIAM J. Comput. 13 (1984): 338–355.

[Ha 71] Harrison, M.C., Implementation of the substring test by hashing, Comm. ACM 14, 12 (1971): 777–779.

[**HR 85**] Hartman, A., and Rodeh, M., Optimal parsing of strings, in: [AG 85]: 155–167.

[**HY 92**] Hashiguchi, K., and Yamada, K., Two recognizable string-matching problems over free partially commutative monoids, Theoret. Comput. Sci. 92 (1992): 77–86.

[**HC 86**] Hébrard, J-J., and Crochemore, M., Calcul de la distance par les sous-mots, R.A.I.R.O. Informatique Théorique 20 (1986): 441–456.

[**He 87**] Held, G., Data Compression Techniques and Applications, Hardware and Software Considerations, John Wiley and Sons, New York, NY, 1987. 2nd edition.

[**Hi 77**] Hirschberg, D.S., Algorithms for the longest common subsequence problem, J. ACM 24 (1977): 664–675.

[**Hi 78**] Hirschberg, D.S., An information theoretic lower bound for the longest common subsequence problem, Inf. Process. Lett. 7 (1978): 40–41.

[**HL 87a**] Hirschberg, D.S., and Larmore, L.L., New applications of failure functions, J. ACM 34 (1987): 616–625.

[**HL 87b**] Hirschberg, D.S., and Larmore, L.L., The set LCS problem, Algorithmica 2 (1987): 91–95.

[**HO 82**] Hoffman, C.M., and O'Donnell. M.J Pattern matching in trees, J. ACM 29,1 (1982): 68–95.

[**HU 79**] Hopcroft, J.E., and Ullman, J.D., Introduction to Automata, Languages and Computations, Addison-Wesley, Reading, Mass., 1979.

[**Ho 80**] Horspool, R.N., Practical fast searching in strings, Software-Practice and Experience 10 (1980): 501–506.

[**HD 84**] Hsu, W.J., and Du, M. W., New algorithms for the LCS problem, J. Comput. Syst. Sci. 29 (1984): 133–152.

[**Hu51**] Huffman, D.A., A method for the construction of minimum redundancy codes, Proceedings of the I.R.E. 40 (1951): 1098–1101.

[**Hu 88**] Hume, A., A tale of two greps, Software—Practice and Experience 18 (1988): 1063–1072.

[**HS 91**] Hume, A., and Sunday, D.M., Fast string searching, Software—Practice and Experience 21, 11 (1991): 1221–1248.

[**HS 77**] Hunt, J.W., and Szymanski, T.G., A fast algorithm for computing longest common subsequences, Comm. ACM 20 (1977): 350–353.

[**IS 93**] Idury, R., and Schäffer, A., Multiple matching of rectangular patterns, in: (STOC'93, 1993): 81–90.

[IS 92] Iliopoulos, C.S., and Smyth, W.F., Optimal algorithms for computing the anonical form of a circular string, Theoret. Comput. Sci. 92 (1992): 87–05.

[Jà 92] Jàjà, J., An Introduction to Parallel Algorithms, Addison-Wesley, Reading, Mass., 1992.

[Ja 92] Jantke, K., Polynomial time inference of general pattern languages, in: (STACS'92, Lecture Notes in Computer Science 166, 1992): 314–325.

[Jo 77] Jones, N.D., A note on linear-time simulation of deterministic two-way pushdown automata, Inf. Process. Lett. 6,4 (1977): 110–112.

[KMR 72] Karp, R.M., Miller, R.E., and Rosenberg, A.L., Rapid identification of repeated patterns in strings, arrays and trces, in: (Proc. 4th ACM Symposium on Theory of Computing, Association for Computing Machinery, New York, 1972): 125–136.

[KR 87] Karp, R.M., and Rabin, M.O., Efficient randomized pattern-matching algorithms, IBM J. Res. Dev. 31 (1987): 249–260.

[K-P 01] Kasai, T., Lee, G., Arimura, H., Arikawa, S., and Park, K., Linear-time longest-common-prefix computation in suffix arrays and its applications, in: (Proc. 12th Combinatorial Pattern Matching, LNCS 2089, Springer-Verlag, 2001) 181–192.

[KLP 89] Kedem, Z.M., Landau, G.M., and Palem, K.V., Optimal parallel suffix-prefix matching algorithm and applications, in: (Proc. ACM Symposium on Parallel Algorithms, Association for Computing Machinery. New York, 1989): 388–398.

[KP 89] Kedem, Z.M., and Palem, K.V., Optimal parallel algorithms for forest and term matching, Theoret. Comput Sci 93,2 (1989): 245–264.

[Kf 88] Kfoury, A.J., A linear-time algorithm to decide whether a word is overlap-free, RAIRO Inform. Théor. Appl. 22 (1988): 135–145.

[KS 99] Kida, T., Shibata, Y., Takeda, M., Shinohara, A., and Arikawa, S., A Unifying Framework for Compressed Pattern Matching, Proc. 6th International Symposium on String Processing and Information Retrieval (SPIRE'99), IEEE Computer Society, pp. 89-96, September 1999.

[KS 92] Kim, J.Y., and Shawe-Taylor, J., An approximate string-matching algorithm, Theoret. Comput. Sci. 92 (1992): 107–117.

[K1 56] Kleene, S.C., Representation of events in nerve nets and finite automata, in: (Shannon and McCarthy editors, Automata studies, Princeton University Press, 1956): 3–40.

[Kn 85] Knuth, D.E., Dynamic Huffman coding, J. Algorithms 6 (1985): 163–180.

[KP71] Knuth, D.E., and Pratt, V.R., Automata theory can be useful, Report, Stanford University, 1971.

[KMP 77] Knuth, D.E., Morris Jr, J.H., and Pratt, V.R., Fast pattern matching in strings, SIAM J. Comput. 6 (1977): 323–350.

[KH 87] Ko, Ker-I, and Hua, Chin-Ming, A note on the two-variable pattern-finding problem, J. Comput. Syst. Sci 34 (1987): 75–86.

[Ko 89] Kosaraju, S.R., Efficient tree-pattern matching, in (FOCS'89, 1989): 178–183.

[LSV 87] Landau, G.M., Schieber, B., and Vishkin, U., Parallel construction of a suffix tree, in: (Automata, Languages and Programming, Lecture Notes in Computer Science 267, Springer-Verlag, Berlin, 1987): 314–325.

[LV 86a] Landau, G.M., and Vishkin, U., Introducing efficient parallelism into approximate string matching, in (STOC, 1986): 220–230.

[LV 86b] Landau, G.M., and Vishkin, U., Efficient string matching with k mismatches, Theoret. Comput Sci 43 (1986): 239–249.

[LV 88] Landau, G.M., and Vishkin, U., Fast string matching with k differences, J. Comput. Syst. Sci 37 (1988): 63–78.

[LV 89] Landau, G.M., and Vishkin, U., Fast parallel and serial approximate string matching, J. Algorithms 10 (1989): 158–169.

[Le 92] Lecroq, T., A variation on the Boyer-Moore algorithm, Theoret. Comput. Sci. 92 (1992): 119–144.

[LZ 76] Lempel, A., and Ziv, J., On the complexity of finite sequences, IEEE Trans. Inform. Theory IT 22, 1 (1976): 75–81.

[Li 81] Liu, L., On string pattern matching: A new model with a polynomial time algorithm, SIAM J.Comput. 10 (1981): 118–139.

[Lo 83] Lothaire, M., Combinatorics on Words, Addison-Wesley, Reading, Mass., 1983. Reprinted by Cambridge University Press, 1997.

[LW 75] Lowrance, R., and Wagner, R.A., An extension of the string-to-string correction problem, J. ACM 25 (1975): 177–183.

[LS 62] Lyndon, R.C., and Schützenberger, M.P., The equation $a^m = b^n c^p$ in a free group, Michigan Math. J. 9 (1962): 422–432.

[Ma 77] Maier, D., The complexity of some problems on subsequences and supersequences, J. ACM 25 (1977): 322–336.

[MS 77] Maier, D., and Storer, J.A., A note on the complexity of superstring problem, Report 233, Computer Science Lab., Princeton University, 1977.

[Ma 89] Main, M.G., Detecting leftmost maximal periodicities, Discrete Applied Math. 25 (1989): 145–153.

[ML 79] Main, M.G., and Lorentz, R.J., An $O(n \log n)$ algorithm for finding repetition in a string, TR CS-79-056, Washington State University, Pullman, 1979.

[ML 84] Main, M.G., and Lorentz, R.J., An $O(n \log n)$ algorithm for finding all repetitions in a string, J. Algorithms (1984): 422–432.

[ML 85] Main, M.G., and Lorentz, R.J., Linear-time recognition of square-free strings, in: [AG 85]: 271–278.

[MR 80] Majster, M.E., and Ryser, A., Efficient on-line construction and correction of position trees, SIAM J. Comput. 9,4 (1980): 785–807.

[Ma 75] Manacher, G., A new linear-time on-line algorithm for finding the smallest initial palindrome of the string, J. ACM 22 (1975): 346–351.

[Ma 76] Manacher, G., An application of pattern matching to a problem in geometrical complexity, Inf. Process. Lett. 5 (1976): 6–7.

[Ma 89] Manber, U., Introduction to Algorithms, Addison-Wesley, Reading, Mass., 1989.

[MM 90] Manber, U., and Myers, E., Suffix arrays: A new method for on-line string searches, in: (Proc. of Ist ACM-SIAM Symposium on Discrete Algorithms, American Mathematical Society, Providence, R.I., 1990): 319–327.

[MP 80] Masek, W.J., and Paterson, M.S., A faster algorithm computing string edit distances, J. Comput. Syst. Sci. 20,1 (1980): 18–31.

[McC 76] McCreight, E.M., A space-economical suffix tree construction algorithm, J. ACM 23, 2 (1976): 262–272.

[Mo 84] Monien, B., Deterministic two-way one-head pushdown automata are very powerful, Inf Process. Lett. 18,3 (1984): 239–242.

[MP 70] Morris Jr, J.H., and Pratt, V.R., A linear pattern-matching algorithm, Report 40, University of California, Berkeley, 1970.

[Mo 68] Morrison, D.R., PATRICIA-practical algorithm to retrieve information coded in alphanumeric, J. ACM 15 (1968): 514–534.

[My 86] Myers, E.W., An $O(ND)$ difference algorithm and its variations, Algorithmica 1 (1986): 251–266.

[MM 89] Myers, E.W., and Miller, W., Approximate matching of regular expressions, Bull. Math. Biol. 51 (1989): 5–37.

[NKY 82] Nakatsu, N., Kambayashi, Y., and Yajima, S., A longest common subsequence algorithm suitable for similar text strings, Acta Informatica 18 (1982): 171–179.

[NW 70] Needleman, S.B., and Wunsch, C.D., A general method applicable to the search for similarities in the aminoacid sequence of two proteins, Journal of Molecular Biology 48 (1970): 443–453.

[Ne 95] Nelson, M., and Gailly, J.-L., The Data Compression Book, M&T Books, New York, 1995.

[NC 92] Néraud, J., and Crochemore, M., A string matching interpretation of the equation $x^m y^n = z^p$, Theoret. Comput. Sci. 92 (1992): 145–164.

[Pe 85] Perrin, D., Words over a partially commutative alphabet, in: [AG 85]: 329–340.

[Pe 90] Perrin, D., Finite automata, in: (Handbook of Theoretical Computer Science, vol B. Formal Models and Semantics, Elsevier, Amsterdam, 1990): 1–57.

[Pl 99] Plandowski, W., Satisfiability of Word Equations with Constants is in PSPACE, in (FOCS 1999): 495–500.

[Qu 92] Quong, R.W., Fast average-case pattern matching by multiplexing sparse tables, Theoret. Comput. Sci. 92 (1992): 165–179.

[RS 59] Rabin, M.O., and Scott, D., Finite automata and their decision problems, IBM J. Research and Development 3 (1959): 114–125. Reprinted in: (Sequential Machines : Selected Papers, Addison-Wesley, Reading, Mass., 1964): 63–91.

[Ra 85] Rabin, M.O., Discovering repetitions in strings, in: [AG 85]: 279–288.

[Ré 89] Régnier, M., Knuth-Morris-Pratt algorithm: An analysis, in: (MFCS'89, Lecture Notes in Computer Science 379, Springer-Verlag, Berlin, 1989): 431–444.

[RR 93] Régnier, M., and Rostami, L., A unifying look at d-dimensional periodicities and space coverings, in: (Combinatorial Pattern Matching, Lecture Notes in Computer Science 684, Springer-Verlag, Berlin, 1993): 215–227.

[RS 85] Restivo, A., and Salemi, S., Some decision results on nonrepetitive words, in: [AG 85]: 289–295.

[Re 92] Revuz, D., Minimization of acyclic deterministic automata in linear time, Theoret. Comput. Sci. 92 (1992): 181–189.

[Ri 77] Rivest, R.L., On the worst-case behavior of string-searching algorithms, SIAM J. Comput. 6, 4 (1977): 669–674.

[RT 85] Robert, Y., and Tchuente, M., A systolic array for the longest common subsequence problem, Inf. Process. Lett. 21 (1885): 191–198.

[RPE 81] Rodeh, M., Pratt, V.R., and Even, S., Linear algorithm for data compression via string matching, J. ACM 28 (1981): 16–24.

[Ry 80] Rytter, W., A correct preprocessing algorithm for Boyer-Moore string searching, SlAM J.Comput. 10 (1980): 509–512.

[Ry 85] Rytter, W., The complexity of two-way pushdown automata and recursive programs, in: [AG 85]: 341–356.

[Ry 86] Rytter, W., The space complexity of the unique decipherability problem, Inf. Process. Lett. 23 (1986): 1–3.

[Ry 88] Rytter, W., On efficient computations of costs of paths of a grid graph, Inf Process. Lett. 29 (1988): 71–74.

[Ry 89] Rytter, W., On the parallel transformations of regular expressions to non-deterministic finite automata, Inf. Process. Lett. 31 (1989): 103–109.

[Ry 00] Rytter, W., Compressed and fully compressed one and two-dimensional pattern matching, proceedings of IEEE, November 2000, Volume 88, Number 11, pp. 1769-1778.

[Ry 02] Rytter, W., On maximal suffixes and constant-space versions of KMP algorithm, in: (LATIN 2002).

[Ry 02a] Rytter, W., Application of Lempel-Ziv factorization to the approximation of grammar-based compression, in: (CPM 2002).

[RD 90] Rytter, W., and Diks, K., On optimal parallel computations for sequences of brackets, in: [Ca 90]: 92–105.

[Sa89] Salton, G., Automatic Text Processing, Addison-Wesley, Reading, Mass., 1989.

[SK 83] Sankoff, D., and Kruskal, J.B., Time Warps, String Edits and Macromolecules: The Theory and Practice of Sequence Comparison, Addison-Wesley, Reading, Mass., 1983. Reprinted by Cambridge University Press, 1999.

[Sc 88] Schaback, R., On the expected sublinearity of the Boyer-Moore string-searching algorithm, SlAM J. Comput. 17 (1988): 648–658.

[SV 88] Schieber, B., and Vishkin, U., On finding lowest common ancestors: simplification and parallelization, SlAM J. Comput. 17 (1988): 1253–1262.

[Se 88] Sedgewick, R., Algorithms, Addison-Wesley, Reading, Mass., 1988. 2nd edition.

[SG 77] Seiferas, J., and Galil, Z., Real-time recognition of substring repetition and reversal, Math. Syst. Theory 11 (1977): 111–146.

[Se 74a] Sellers, P.H., An algorithm for the distance between two finite sequences, J. Comb. Th A 16 (1974): 253–258.

[Se 74b] Sellers, P.H., On the theory and computation of evolutionary distances, SIAM J. Appl Math. 26 (1974): 787–793.

[Se 80] Sellers, RH., The theory and computation of evolutionary distances: Pattern recognition, J. Algorithms 1 (1980): 359–373.

[Se 85] Semba, I., An efficient string-searching algorithm, J. Inform. Process. 8 (1985): 101–109.

[Sh 81] Shiloach, Y., Fast canonization of circular strings, J. Algorithms 2 (1981): 107–121.

[SV 81] Shiloach, Y., and Vishkin, U., Finding a maximum, merging and sorting in parallel computation model, J. Algorithms 2 (1981): 88–102.

[S1 83] Slisenko, A.O., Detection of periodicities and string matching in real time, J. Sov. Math. 22, 3 (1983): 1326–1387.

[Sm 82] Smit, G. de V., A comparison of three string-matching algorithms, Software-Practice and Experience 12 (1982): 57–66.

[Sm.90] Smith, RD., An Introduction to Text Processing, The MIT Press, Cambridge, Mass., 1990.

[Sm.02] Smyth W.F., Computing patterns in strings, Addison-Wesley Longman, 2002, to appear.

[Sp 86] Spehner, J.-C., La reconaissance des facteurs d'un mot dans un texte, Theoret. Comput. Sci. 48 (1986): 35–52.

[St 94] Stephen G.A., String searching algorithms, World Scientific Press (1994).

[St 77] Storer, J.A., NP-completeness results concerning data compression, Report 234, Princeton University, 1977.

[SS 78] Storer, J.A. , and Szymanski, T. G. , The macro model for data compression, in: (Proc. 10th ACM Symposium on Theory of Computing, Association for Computing Machinery, New York, 1978): 30–39.

[St 88] Storer, J.A., Data Compression: Methods and Theory, Computer Science Press, Rockville, MD, 1988.

[SG 98] Stoye, J., and Gusfield, D., Simple and Flexible Detection of Contiguous Repeats Using a Suffix Tree, in: (Proc. 9th Combinatorial Pattern Matching, 1998) 140–152.

[Su 90] Sunday, D.M., A very fast substring search algorithm, Comm. ACM 33, 8 (1990): 132–142.

[**Tu 88**] Tarhio, J., and Ukkonen, E., A greedy approximation algorithm for constructing shortest common superstrings, Theoret. Comput. Sci. 57 (1988): 131–146.

[**TU 90**] Tarhio, J., and Ukkonen, E., Boyer-Moore approach to approximate string matching, in: (Proc. 2nd Scandinavian Workshop in Algorithmic Theory, Lecture Notes in Computer Science 447, Springer-Verlag, Berlin, 1990): 348–359.

[**Th68**] Thompson, K., Regular expression search algorithm, Comm. ACM 11 (1968): 419–422.

[**Ti 84**] Tichy, W.F., The string-to-string correction problem with block moves, ACM Trans. Comput. Syst. 2 (1984): 309–321.

[**Uk 85a**] Ukkonen, E., Finding approximate patterns in strings, J. Algorithms 6 (1985): 132–137.

[**Uk 85b**] Ukkonen, E., Algorithms for approximate string matching, Information and Control 64 (1985) 100–118.

[**Uk 92**] Ukkonen, E., Approximate string matching with q-grams and maximal matches, Theoret. Comput. Sci. 92 (1992): 191–211.

[**Uk 92**] Ukkonen, E., Constructing suffix trees on-line in linear time, in: (IFIP'92): 484–492.

[**UW 93**] Ukkonen, E., and Wood, D., Approximate string matching with suffix automata, Algorithmica 10,5 (1993): 353–364.

[**Vi 85**] Vishkin, U., Optimal parallel pattern matching in strings, Information and Control 67 (1985): 91–113.

[**Vi 91**] Vishkin, U., Deterministic sampling, a new technique for fast pattern matching, SIAM J. Comput. 20, 1 (1991): 22–40.

[**Vi 87**] Vitter, J.S., Design and analysis of dynamic Huffman codes, J. ACM 34 (1987): 825–845.

[**Wa 74**] Wagner, R.A., Order-n correction for regular languages, Comm. ACM 17, 6 (1974): 265–268.

[**Wa 75**] Wagner, R.A., On the complexity of extended string-to-string correction problem, in: (Proc. 7th ACM Symposium on Theory of Computing. Association for Computing Machinery, New York, 1975): 218–223.

[**WF 74**] Wagner, R. A. , and Fischer, M.J., The string-to-string correction problem, J. ACM 21 (1974): 168–178.

[**Wa 89**] Waterman, M.S., Mathematical Methods for DNA Sequences, CRC Press, Boca Raton, Fla., 1989.

[We 73] Weiner, P., Linear pattern matching algorithms, in: (Proc. 14th IEEE Annual Symposium on Switching and Automata Theory, Washington, DC, 1973): 1–11.

[We 84] Welch, T.A., A technique for high-performance data compression, IEEE Computer 17,6 (1984): 8–19.

[WMB 99] Witten, Ian H., Moffat, Alistair, and Bell, Timothy C., Managing Gigabytes: Compressing and Indexing Documents and Images, Morgan Kaufmann Publishing, 1999.

[WNC 87] Witten, I.H., Neal, R.M., and Cleary, J.G., Arithmetlc coding for data compression, Commun. ACM 30, 6 (1987): 520–540.

[WC 76] Wong, C.K. , and Chandra, A. K., Bounds for the string-editing problem, J. ACM 23 (1976): 13–16.

[WM 92] Wu, S., and Manber, U., Fast text searching allowing errors, Comm. ACM 35,10 (1992): 83–91.

[Ya 79] Yao, A.C., The complexity of pattern matching for a random string, SIAM J. Comput. 8 (1979): 368–387.

[Zi 92] Zipstein, M., Data compression with factor automata, Theoret. Comput. Sci. 92 (1992): 213–221.

[ZC 89] Zipstein, M., and Crochemore, M., Transducteurs arithmétiques, Rapport L.I.T.P 89–12, Université Paris 7, 1989.

[ZL 77] Ziv, J., and Lempel, A., A universal algorithm for sequential data compression, IEEE Trans. Inform. Theory 23 (1977): 337–343.

[ZL 78] Ziv, J., and Lempel, A., Compression of individual sequences via variable length coding, IEEE Trans. Inform. Theory 24 (1978): 530–536.

[ZP92] Zwick, U., and Paterson, M.S., Lower bounds for string-matching in the sequential comparison model, manuscript, 1992.

Index

www.ingramcontent.com/pod-product-compliance
Lightning Source LLC
LaVergne TN
LVHW022335060326
832902LV00022B/4043